THE CHEATING CELL

The Cheating Cell

How Evolution Helps Us Understand and Treat Cancer

Athena Aktipis

PRINCETON UNIVERSITY PRESS

PRINCETON AND OXFORD

Published by Princeton University Press
41 William Street, Princeton, New Jersey 08540
6 Oxford Street, Woodstock, Oxfordshire OX20 1TR

press.princeton.edu

All Rights Reserved
First paperback printing, 2021
Paperback ISBN 978-0-691-21219-7
Cloth ISBN 978-0-691-16384-0
ISBN (e-book) 978-0-691-18608-5

Library of Congress Control Number: 2019954861

British Library Cataloging-in-Publication Data is available

Editorial: Alison Kalett and Abigail Johnson
Production Editorial: Karen Carter
Production: Jacqueline Poirier
Publicity: Sara Henning-Stout and Katie Lewis

Cover design and illustrations by Sukutangan

This book has been composed in Adobe Text and Gotham

To all the beautiful monsters who came before us.

CONTENTS

ACKNOWLEDGMENTS

This book is the product of many late-night conversations at the kitchen table, lunches in poorly lit conference rooms, coffee breaks in the basement, happy hours on the porch, and back-of-the-room scribbling at academic meetings while listening to excellent talks from my brilliant colleagues. I am grateful to the many collaborators, colleagues, and friends who have shared their brains with me over the decades, contributing to the ideas that form the basis for this book. It would be impossible to thank each of them here—I have had literally hundreds of conversations that have shaped my thinking and the ideas in this book, and I am grateful to everyone who has shared their thoughts and ideas with me. You know who you are. Thank you. And please accept my apologies for not being able to acknowledge every one of you by name here.

I am especially grateful to my colleagues, friends, and students who have taken the time to read drafts of this book and give me feedback: Jessica Ayers, David Buss, Lee Cronk, Pauline Davies, Mark Flinn, Rick Grosberg, Michael Hechter, Steffi Kapsetaki, Joseph Mamola, Pranav Menon, Anya Plutynski, Pamela Winfrey, and all of the students in Carlo Maley's spring 2018 Cancer Evolution class. And a very special thank-you to Andrew Read for not only reading the manuscript but also providing extraordinarily detailed, thoughtful, and valuable comments. Thank you also to Zachary Shaffer, Bob Gatenby, Amy Boddy, and many other colleagues who corresponded with me during the writing of this book, answering questions and providing useful suggestions.

Thank you to the team at Princeton University Press, especially my editor, Alison Kalett, who provided a perfectly motivating balance of encouragement and honest feedback. Thank you also to my scientific editor, Jane Hu, who provided essential comments, editorial advice, and support during the most challenging phases of writing this book. I am also greatly indebted to Amanda Moon, whose keen editorial eye greatly improved the final version of this manuscript. Thank you to two anonymous reviewers and one not-so-anonymous reviewer, James DeGregori, for thoughtful and detailed comments on the manuscript. Also, thank you to my dedicated research assistant, Nicole Hudson, who worked tirelessly to compile the endnotes, finalize the formatting, and acquire permissions for several of the images used in this book. And a very special thank-you to my lab manager, Cristina Baciu, for much help along the way with all steps of the book, from the initial research to the final formatting, and especially for taking such good care of my lab while my attention was on this book. Cristina, thank you for your dedication, support, and most especially for your generosity of mind and spirit. I was also very lucky to have the talented Alex Cagan illustrating this book. Thank you, Alex, for your attention to detail, your patience, and your tolerance of my (admittedly) picky graphic design preferences.

I would not have been able to write this book without the support of many universities, institutions, and other organizations where I spent time during the writing. This book was conceived during a wonderful year at the Wissenschaftskolleg zu Berlin (the Institute for Advanced Study in Berlin) in the Cancer Evolution working group. Thank you to my colleagues in that group, and all the other 2013–14 fellows who provided a truly unparalleled intellectual environment, especially the late Paul Robertson. Thank you, Paul, for all the wonderful breakfasts discussing many of the ideas that came to form this book; you were both a dear friend and a most valued colleague. I am also most grateful to the members of the International Society for Evolution, Ecology and Cancer for providing a wonderful environment for the development of many of the ideas that appear in this book. Also, a heartfelt thanks to the crews at Berdena's and Firecreek Coffee, who kept me perfectly caffeinated through many drafts and revisions.

I wrote most of this book while working as an assistant professor in the Department of Psychology at Arizona State University. I am grateful to my colleagues in my department and across ASU for supporting my writing of this book. Thank you especially to the former chair of my department, Keith Crnic, and the current chair, Steve Neuberg, for supporting me in writing and, more generally, for supporting my interdisciplinary research program. Thank you also to the president of ASU, Michael Crow, for his support and also for creating an interdisciplinary environment that I have benefited from in more ways than I can fathom.

I am also extremely lucky to have had many teachers and mentors over the decades of my life who taught me, guided me, and advised me on my interdisciplinary academic trajectory. Thank you to my teachers at Willowbrook High School, especially Will Nifong (who showed me the joys of language), Vicky Edwards (who first taught me how to write), Joy Joyce (who showed me how important economics is in everything), John Mostacci (who inspired both scholarly inquiry and irreverence), and Ed Raddatz (who showed me how academic learning can and should apply to real problems). Thank you to my professors at Reed College, especially my advisors Allen Neuringer (who showed me the importance of not getting stuck in a rut) and Noel Netusil (who found me wandering the halls in a daze after a horrible meeting with my assigned freshman advisor and generously "adopted" me). I was lucky, also, to have many dedicated professors in graduate school at the University of Pennsylvania, including my advisor, Rob Kurzban, the late John Sabini (who provided much guidance), and Sharon Thompson-Schill (who provided mentorship, friendship, and encouragement when it was most needed). Thank you also to the professors in the Department of Ecology and Evolutionary Biology at the University of Arizona, where I did my postdoc, especially Rick Michod and Aurora Nedelcu. A very special thanks to my postdoc advisor—now friend and colleague—John Pepper, for first sparking my interest in evolution and cancer and then fanning the flames. Thank you also to the many colleagues and friends who have supported me during the writing of this book: Martie Haselton (for constant words of encouragement), Nicole

Hess (for unwavering support in all matters large and small), Sarah Hill (for always being willing to entertain wildest of ideas), and Barbara Natterson-Horowitz (for helping me stay focused on the big picture no matter what distractions crossed my path).

Thank you also to my colleague Mel Greaves for providing me with the best piece of advice I've ever gotten during a dinner conversation: "Be careful whose advice you take." It is perhaps the only advice that I consistently follow. (In the same conversation he gave me the advice to eat less salt. Sorry, Mel, I'm not giving up my salt.)

I am forever indebted to the babysitters and nannies who helped take care of my children during the years it took to write this book. Thank you especially to Veronica Mata Ford and Lisa Lessard for putting your hearts and souls into our family.

I am ever grateful to my parents, Stelios Aktipis and Helga Fitz Aktipis. Dad, thank you for reassuring me throughout grade school that if I just stuck it out until college I would start to enjoy school. Mom, thank you for teaching me to always look at the world from multiple perspectives and to be intellectually fearless. I still miss you every day, but I carry you with me in my heart—and probably also in my breast, brain, thyroid, and immune system, thanks to the microchimeric cells you passed to me while I was in your womb.

The most important thanks for this book goes to my friend, colleague, and husband, Carlo Maley. Thank you, Carlo, for the many late-night conversations, the patient proofreading, the rapid responses to various queries and—most important—for taking such good care of our children during the many weekends I spent writing this book. And thank you to my children, Avanna, Monty, and Vaughn, for your love, support, and understanding while I was writing this book. There was no part of writing this book that was more rewarding than getting to talk with all of you about the ideas, the process of writing, and my hopes for it. So thank you, my dear children, for your interest in—and patience with—this somewhat unruly brainchild that's been a part of our family for the last seven years.

THE CHEATING CELL

1

Introduction

EVOLUTION IN THE FLESH

This is a book about cancer: its ancient origins, its modern manifestations, and its future fate. It is a book about where cancer came from, why it exists, and why it is so hard to cure.

This is also a book about a new way of looking at cancer—not as something that must be eliminated at all costs, but rather as something that must be controlled and shaped into a companion that we can live with.

Life has struggled with cancer since the dawn of multicellularity about two billion years ago. When we think of life on Earth, we typically think of multicellular organisms like animals and plants that are made up of more than one cell. The cells in a multicellular organism essentially divide the labor of making a living, cooperating, and coordinating to do all the functions that are needed in the body. On the other hand, unicellular life forms—like bacteria, yeasts, and protists—are made of a single cell that does all of the jobs of keeping that cell alive. Unicellular life dominated our planet for billions of years before multicellular life gained an evolutionary foothold. The world was cancer-free during these two billion years when unicellular life reigned. But when multicellular life entered the scene, it ushered in a new player: cancer.

Cancer is a part of us, and it has been since our very beginnings as multicellular organisms. Remnants of cancers have been found in the skeletons of ancient humans, from Egyptian mummies to Central and South American hunter-gatherers. Cancer has been found in 1.7-million-year-old bones of our early human ancestors in "the cradle of humankind" in South Africa. Fossil evidence of cancer goes back further still; it is found in bones tens and even hundreds of millions of years old, from mammals, fish, and birds. Cancer goes back as far as the days when dinosaurs dominated life on our planet, and back even further than that, to a time when life was microscopically small. Cancer began before most of life as we know it even existed.

In order to manage cancer effectively, we must understand the evolutionary and ecological dynamics that underlie it. But we must also change our way of thinking about cancer, from viewing it as a temporary and tractable challenge to seeing it as a part of who we are as multicellular beings. Before multicellular life evolved, cancer did not exist because there were no bodies for cancer cells to proliferate inside of and ultimately invade. But once multicellular life emerged, cancer was able to emerge as well. Our very existence as multicellular organisms—as paragons of multicellular cooperation—is inextricably tied to our susceptibility to cancer.

In this book we will see how our bodies are made of cells that cooperate in myriad ways to make us functional multicellular organisms—for example, by controlling cell proliferation, distributing resources to cells that need them, and building complex organs and tissues. We will also see how cancer can evolve to exploit the cooperative cellular nature of our bodies: proliferating out of control, exploiting the resources in our bodies, and even turning our tissues into specialized niches for their own survival. In a word, cancer is cheating in the game that forms the most fundamental foundations of multicellular life.

A better understanding of the essential nature of cancer can help us to prevent and treat it more effectively, and also help us to see that we are not alone in our struggles with cancer. All forms of multicellular life are affected by cancer. Our evolutionary relationship with cancer has shaped who we are. And if we want to truly understand what cancer is, we must understand how it evolved and how we evolved along with it.

We can look to the natural world to recognize what cancer is and how it evolves. One of the most beautiful examples is the crested cactus. Sometimes the cells in the growing tip of a cactus will mutate as a result of damage or infection. These mutations can disrupt the normal controls on cell proliferation during the growth of the plant. This often leads to striking formations: desert saguaros that look like they are wearing crowns, potted cacti that look like brains, garden cacti with knobby geometrical surfaces that evoke modern art (figure 1.1). Crested cacti are highly prized by professional botanists and backyard cactus lovers alike due to their beautiful and unusual mutated forms.

When I first saw a crested cactus on a visit to Arizona years ago, I was fascinated by the beauty and geometry of the plant. When I returned to my hotel room, I spent several hours looking at photographs of these natural biological formations and reading about them. I learned that the disrupted growth patterns of the mutated crested cacti sometimes result from damage during storms, sometimes from bacteria or viruses, and sometimes from genetic mutations during development.

I also learned that mutations that disrupt plant growth patterns are not unique to cacti—they happen across many plant species, from dandelions to pine trees. The technical term for these disrupted growth formations in plants is *fasciation*. Fasciated plants are often more delicate than their nonfasciated cousins, and sometimes they do not flower normally, making it harder for them to reproduce and propagate themselves—however, fasciated plants are often cared for and propagated by gardeners and botanists. With proper care, crested cacti and other fasciated plants can live for decades with these cancer-like formations.

Learning about crested cacti marked the beginning of my fascination with cancer across forms of life. I thought to myself at the

FIGURE 1.1 Cacti can develop abnormal growths as a result of disruption to the normal growth pattern. This can result in many beautiful and unique growth patterns that have similarities to cancer in animals. These cancer-like phenomena in plants, known as fasciations, can have negative effects on plant fitness including less flowering and greater susceptibility to injury or disease, but these plants also can survive with these cancer-like forms for decades if they have proper care. Images from left to right are a crested saguaro cactus, *Carnegiea gigantea*; a "brain cactus," *Mammillaria elongata cristata*; a "totem pole cactus," *Pachycereus schottii* f. *monstrosus*; and a *Cereus jamacaru* f. *cristatus*.

time: if we are going to understand cancer—what it is and why it can threaten our well-being and our lives—surely we need to know where cancer comes from, which means understanding the evolutionary origins of cancer across the tree of life. As I continued on my journey to understand the evolutionary origins of cancer, I discovered that cancer and cancer-like formations are ubiquitous across multicellular life. I found that cacti were not alone in having these cancer-like formations, but that there were also a myriad of other organisms that had cancer-like growths. I found pictures of mushrooms and coral and algae and insects with cancer-like growths. And I discovered that cancer was common across animals—from wild animals, to animals kept in zoos, to the domesticated animals that live with us in our own homes.

Why, I wondered, was cancer so pervasive across all forms of multicellular life? Cancer is uniquely a problem of multicellularity because multicellular life is made of many cells—cells that usually cooperate and regulate their behavior to make us functional organisms. Unicellular life forms don't get cancer because they are made of just one cell. This means that, for unicellular life, cell proliferation is the same as reproduction. But for multicellular life, too much cell proliferation can disrupt the normal development and structure of the multicellular organism.

You might feel like a unitary being, but in reality you are made of trillions of cells that are cooperating and coordinating their behavior every millisecond to make you a functional human being. The number of cells inside our bodies is mind-boggling—more than four thousand times the number of humans on Earth. We are thirty trillion cooperating, evolving, consuming, computing, gene-expressing, protein-producing cells. The body is literally a world unto itself. Each of these cells is like a little homunculus inside you, taking information from its environment, processing that information using complex genetic networks, and changing what it does in response to those inputs. Each cell has its own set of genes, unique gene expression (i.e., the specific proteins the cell is making) and its own physiology and behavior. The cooperation happening inside us is quite astounding. How can thirty trillion cells make a being that seems so much like one single entity with one set of goals? How can I be made of so many cells yet feel so unitary?

One answer to these questions comes from evolutionary biology: We act and feel like unitary organisms because evolution has shaped us to be cooperative cellular societies. Perhaps we feel like unitary beings because evolution has fashioned us to act as though we are. We have been shaped by nearly one billion years of evolution on multicellular bodies to have cells that act in a way that enhances the survival and reproduction of the cooperative cellular society as a whole—the multicellular body. Our cells constrain their proliferation, divide labor, regulate their resource use, and even commit cell suicide for the benefit of the organism. The scope of cooperation inside us is beyond anything humans have ever accomplished—the cells inside us behave like a success story of a utopia, sharing resources, taking care of the shared environment, and regulating their behavior for the good of the body.

But sometimes this cellular cooperation breaks down. And when it does, this can set off an evolutionary and ecological process in the body that culminates in the ultimate form of cellular cheating: cancer. Cancer is what happens when cells stop cooperating and coordinating for the benefit of the multicellular body and start overusing resources, trashing the shared environment of the body, and replicating out of control. Inside the body, these cheating cells can have an evolutionary advantage over normal cells, despite the fact that they can damage the health and survival prospects of the body of which they are a part.

Although we feel like unitary individuals, fundamentally we are not. Evolution fashioned us to be incredibly functional as multicellular organisms, but we cannot escape the fact that we are a population of cells. Because we are made of a vast population of cells, evolution naturally occurs within our bodies. Cells in the body can evolve just as organisms in the natural world evolve. This is a very different way of thinking about who we are. In the traditional view, we are a unitary and relatively static "self." However, not only are we made of trillions of individual cells, we are made of trillions of cells that are part of a constantly evolving population. We are not one entity, but rather many entities. And as we age, the population of cells that composes us continues to evolve, often in directions that put us at risk for cancer.

Cells are of course a part of who we are, yet they are also very much their own entities. Cells express genes, they process information, they behave—moving, consuming resources, and building extracellular structures like tissue architecture. In addition, they are a population inside our body that is evolving in a complex ecological environment. We need both of these perspectives—cells as a part of us and cells as their own unique evolving entities inside of us—to understand what cancer is and why we are vulnerable to it.

From the perspective of our bodies, cancer is a threat to our survival and well-being. From the perspective of the cell, cancer cells are only doing what every other living thing on this planet does: evolving in response to the ecological conditions they are in, sometimes in ways that are detrimental to the system of which they are a part. This leads to a seemingly paradoxical evolutionary scenario: Evolution favors bodies that are good at suppressing cancer, but evolution also favors cells inside the body that have the characteristics of cancer cells, such as rapid proliferation and high metabolism. How can both of these facts be true—on the one hand evolution favors cancer cells, while on the other hand evolution favors cancer suppression? As the narrative in this book unfolds, I will reveal how an evolutionary perspective can help us understand this apparent paradox.

The scale of cellular cooperation in our bodies is astonishing. But even more stunning is how resilient our bodies can be when faced with cellular cheating—how we can survive and thrive despite the threat of cancer. Multicellular bodies have evolved many different cancer suppression mechanisms over billions of years. These cancer suppression systems allow us to keep cellular cheating under control. Looking across species, we can also witness the diversity and power of these cancer suppression systems at work, and gain insights and inspiration for how we might better treat cancer in human beings. Like the crested cacti, which can coexist with their cancer-like growths for decades, perhaps we also can live with cancer.

Before I learned about cancer's evolutionary nature, I thought of cancer as nothing more than a rather uninteresting disease. My work focused on deep and fundamental questions about the evolution of life: Why are so many organisms social? What makes cooperation stable despite the possibility of exploitation from so-called cheaters? I had always been drawn to theoretical questions, so I shied away from any topic that seemed to require the memorization of an endless catalog of facts with no framework to hold them together. Cancer seemed to be one of those topics—no theoretical grounding, just an endless number of studies on mechanism after mechanism with no underlying principles to discover. It was certainly worthy of study because of its importance to human health, but I had no interest in studying it myself.

Then I moved to the University of Arizona to work as a postdoctoral researcher, and I began working with John Pepper, a pioneer in cancer evolution, a new field at the time. I realized that cancer was a cellular example of exactly what I was already studying: the challenges of maintaining cooperation in large-scale evolving systems in the face of cheaters.

My view of cancer started to change. I realized that cancer is a living entity evolving rapidly in the ecosystems of our bodies. It follows the same rules that all evolutionary and ecological systems follow. Placing cancer in an evolutionary framework provided a starting point for understanding its complexity.

The great evolutionary biologist Theodosius Dobzhansky—one of the pioneers of evolutionary thinking in the twentieth century—once said, "Nothing in biology makes sense except in the light of evolution." As I came to see cancer in this evolutionary light, I realized that cancer biology hadn't made sense to me before that point because I hadn't applied an evolutionary and ecological lens to understand it.

If Dobzhansky were around today, he might very well say, "Nothing in cancer biology makes sense except in the light of evolution." Evolution, ecology, and cooperation theory offer a starting point for understanding why cancer is such a complex, powerful, and dynamic force, and they can help us better understand who we are. And these same tools can help us understand how cancer has shaped—and continues to shape—all of multicellular life.

Evolutionary theory explains how cancer can exist on two different levels. First, it shows how evolution among the cells in our body—often called somatic evolution—leads to cancer. Cancer is the literal embodiment of evolution: cells in our bodies are evolving inside us. The cells in our body vary in terms of how evolutionarily fit they are inside our bodies; some cells replicate faster and survive longer than others. The cells that proliferate more and survive longer subsequently make up a larger portion of the next generation and eventually come to dominate the population. This is evolution by natural selection, the same process that has shaped the evolution of organisms in the natural world.

In addition, evolutionary theory helps explain why cancer has persisted over the course of life on Earth. Organisms have evolved over millions of years to suppress cancer—to keep somatic evolution under control—so that we can live long and evolutionarily successful lives. These cancer suppression systems are the reason that multicellular life is even possible—without them, multicellularity would never have been able to overcome the challenges of cellular cheating from within. But these cancer suppression systems are not perfect. Evolutionarily speaking, keeping would-be cancer cells 100 percent under control is not possible.

The reasons we can't completely suppress cancer are varied, and each is fascinating in its own right. For example, one reason why organisms don't evolve to suppress cancer completely is because of trade-offs with other traits that affect the fitness of the organism, like fertility. Sometimes, lower cancer risk is associated with lower fertility, creating an evolutionary bind for organisms evolving to suppress cancer. In addition, organisms can't suppress cancer completely because there is a mismatch between past and current environments: modern humans are exposed to mutagens like cigarette smoke and lifestyle factors such as low physical activity, which lead to greater susceptibility to cancer. An even more bizarre reason why we can't suppress cancer is that there is a battle over our growth happening between genes we inherit from our mothers and genes we inherit from our fathers. Some of the genes we inherited from our fathers are epigenetically set to promote growth and cell proliferation, contributing to an increased risk of cancer. Cancer exists because of the tension between two evolutionary processes acting on two different scales: cells evolve in the body via somatic evolution, and bodies can't evolve to completely suppress this process of somatic evolution.

The environment that cancer cells occupy in the body can greatly affect whether would-be cancer cells die or survive and thrive. In cancer biology, the environment around a tumor is called the tumor microenvironment. This is essentially the ecosystem of the tumor, in many ways like an ecosystem in the natural world. The ecosystem of the tumor provides necessary resources that allow the cells in the tumor to survive and thrive, but it can also threaten the survival of cells when resources run out, waste products build up, and the immune system starts preying on cancer cells. Cancer cells can alter their environment as they consume resources such as glucose; for instance, they can reduce the supply of resources for neighboring cells and leave behind waste products like acid. Those changes, however, can trash the ecology of the cancer cells, making it difficult for them to survive and thrive. The destruction of the microenvironment can create selection pressures for those cells to move. Cells that can move and are able to relocate to a new and better environment in the body will survive and leave more cellular progeny, spurring the evolution of invasive and metastatic cells. Ecology is central to the ways in which cancer emerges and progresses. Just as we can't understand how and why organisms evolve without knowing about their environments, we can't understand how and why cancer evolves without knowing the ecological dynamics in and around a cancerous tumor.

The common metaphor for cancer is war—patients "fight," "battle," "win," or "lose." The war metaphor for cancer is powerful and compelling. It can help to rally support for cancer research and bring people together around a common goal, but it can also be misleading. We can't completely eradicate something that is fundamentally a part of us. This aggressive approach to the disease seems like a good idea if we view cancer as an enemy to eradicate. But unless we see cancer for what it is—a population of diverse cells evolving in response to every treatment we throw at it—we run the risk of discounting or outright dismissing less aggressive forms of treatment.

The war metaphor encourages an aggressive outlook that can lead to other consequences as well. When we treat cancer with high-dose therapies, this can give an evolutionary advantage to cells that are resistant to those therapies, making our treatments less effective for long-term tumor control. For terminal cancers, attacking with the highest-dose therapies is often not the ideal strategy. Approaching cancer with an aggressive mind-set can also have a negative effect on prevention. When people are presented with a war metaphor for cancer, they report being less likely to engage in some cancer-prevention behaviors, like stopping smoking. In addition, aggressive language related to treatment can increase stress levels for cancer patients and their families.

Cancer is not an enemy in the typical sense of the term. Cancer is not an organized and homogenous army that is collectively set on our destruction. Instead, it is a disorganized and heterogenous population of cells that is dynamically responding to our treatments. When we fight cancer, we are fighting against an inevitable process: the process of evolution. We can slow down that process or change its course, but we can't make the process stop.

Cancer is the literal embodiment of evolution. It is evolution in the flesh. We are susceptible to cancer because we are made of a population of cells that evolves over our lifetimes. Cancer will be here for as long as multicellular life endures on our planet. The sooner we can accept that, the sooner we can use our knowledge to effectively keep cancer under control.

We can't win a war against a process of evolution, a process of ecological change in our bodies, a process of cellular free-riding taking over multicellular cooperation. But we can shape that process so it is less harmful to us. We can gain insights and strategies to help us shape cancer into something more benign and less threatening—in other words, something we can live with.

Fighting a war on cancer for which the only acceptable outcome is complete destruction of the enemy, as opposed to using circumscribed strategies to exploit cancer's vulnerabilities, is much like the contrasting war strategies of two Greek gods of war, Athena and Ares. I grew up in a largely Greek household, living first in Athens and then in the suburbs of Chicago, and Greek mythology was an important part of my childhood. I was raised in part by my grandmother, Athena (after whom I was named). Of course, I had an interest in understanding my namesake. Athena is the goddess of wisdom and war, but not just any kind of war; she is the goddess of strategy. Rather than winning by brute force, Athena wins by understanding the goals and vulnerabilities of the enemy and then exploiting those to achieve victory with minimum force and without unnecessary collateral damage. Ares, on the other hand, approaches battle with maximum aggression and with the goal of inflicting as much damage as possible on the enemy at all costs.

Which approach is right for cancer? Should we fight by brute force like Ares, or should we plan a strategy (like Athena would do) that exploits the vulnerabilities of our adversary? From what we know about cancer, it is clear that Athena's approach is more likely to extend the lives of cancer patients, improving their quality of life as well. (And I'm not just saying that because I'm named after her.)

Cancer is an inexorable part of our lives and histories as individuals, and as multicellular organisms. In this book, I will use our evolutionary history as the basis for discovering insights into what cancer is, why it emerges, and how we can better treat it. I will argue that cancer is more than just a disease; it is a window into the origins of life, the challenges of large-scale cooperation, the nature of multicellularity, and the process of evolution itself.

2

Why Does Cancer Evolve?

Let's play a little game. Which of these statements about cancer is true?

1. We get cancer because it helps to control human population size.
2. We get cancer because it sends a message that we need to take better care of ourselves.
3. We get cancer because it is better for our offspring if we die at an earlier age before we become a burden.
4. We get cancer because cells in our bodies that survive and proliferate quickly leave more cellular descendants.

All of these statements may sound plausible, but only the last one is true. Notice that each statement seeks to explain why we get cancer at a different level: the first posits that cancer evolves to benefit our species, the second that it benefits us as individual humans, the third that it benefits our kin, and the fourth that cancer evolves to benefit individual cells.

When cells rapidly multiply and survive when they shouldn't, the process can have devastating effects on the organism. Our bodies are built to be bastions of cellular cooperation, with systems for sharing resources and building an environment inside the body that allows cells to survive, thrive, and do all the jobs necessary to make our bodies function. Cancer cells exploit that cooperation, getting an evolutionary leg up on their normal neighbors. This process is essentially the same as the process of evolution that happens among organisms in the natural world (with some exceptions, which I will come back to later).

Evolution is defined as a change in gene frequencies in a population over time. With our bodies, the frequency of different genetic mutations among our cells changes every time a cell divides or dies. Cells, just like organisms, compete with one another for survival and reproductive opportunities. The cells that are most fit in the environment of our bodies (i.e., those that survive and replicate best) end up growing in the cellular population that makes up our multicellular bodies. Unfortunately, one of the ways that cells can get an evolutionary leg up on their cellular neighbors is by ignoring the restrictions on cell proliferation and survival that otherwise keep us cancer-free.

The idea that cancer is an evolutionary process goes back many decades. In the 1970s, cancer researcher Peter Nowell described cancer as an evolutionary process based on the accumulation of genetic mutations. In the same decade, the physician-cum-molecular-biologist John Cairns pointed out that our bodies are likely to have protective mechanisms that help to keep cancer from evolving within us. The roots of this idea go back even further, to the idea that competition happens within the body during development (proposed by Wilhelm Roux in the late 1800s), to the notion that cellular mutations can lead to "egoistic" cell behavior (proposed by Theodor Boveri in the early 1900s), and to the idea of the stepwise progression of cancer (proposed by Leslie Foulds in the mid-1900s). In the late 1990s and early 2000s, the field began to grow as researchers including cancer biologist Mel Greaves, evolutionary geneticist Leonard Nunney, computational evolutionary biologist Carlo Maley, and many others began to study cancer using techniques and frameworks from evolutionary biology.

I had the privilege of coming into the field of cancer evolution in the mid-2000s, just as it was experiencing tremendous growth.

This growth came as a result of new tools and methods in genomics that allowed researchers to look at the evolutionary dynamics in tumors and follow lineages of cancer cells (called clonal expansions) during cancer progression. A clonal expansion refers to a group of cells that came from a common ancestor and share a mutation. As we know, cells that proliferate and thrive in the environment of the body end up leaving more cellular descendants, which gives rise to clonal expansions. Some clonal expansions are made up of cells that have a proliferation or survival advantage over their neighbors (a process evolutionary biologists call natural selection). Other clonal expansions can be the result of random processes (what evolutionary biologists refer to as genetic drift).

In the case of cells in our bodies, mutations that allow cells to divide rapidly and survive better will increase in frequency in the population of cells through natural selection, setting the stage for cancer. Natural selection is the process of change in the characteristics of a population over time as a result of the differential reproduction and survival of individuals within the population that have different traits. But evolution can also happen as a result of random influences on cell success. Genetic drift occurs in small populations—of organisms, cells, or any living things—when random chance leads some individuals to survive and reproduce better than others. Traits that have no effect on survival or reproduction can spread or decline in a population simply as a result of random chance. Both of these processes—natural selection and genetic drift—have a role in how cancer evolves inside us. For the purposes of this chapter, and the book more generally, I will focus on natural selection because it helps us explain one of the central evolutionary paradoxes of cancer: how cancer cells can evolve to destroy the very hosts that they depend on for their own survival.

How Cancer Cells Evolve Inside Us

Our bodies are vast worlds in which populations of cancer cells can evolve. We are used to thinking about evolution as a glacially slow process: over thousands of years, random variations happen and are sometimes advantageous to the organisms possessing them, slowly transforming a population of individuals as they evolve to better fit within their environment. Because we often think about evolution as a slow process, it can be difficult to wrap our heads around the way evolution works among cells inside our bodies. If we think of evolution as an inherently slow process, how could it possibly work fast enough to select for cancer cells over the course of a human lifetime? The answer is that the timescale of evolution is completely different inside the body: the generation time of a cancer cell (in other words, the time it takes to reproduce) is extremely short—often about one day—and the population sizes are in the billions, so large that the pace of evolution can be extremely fast. In fact, more cellular evolution happens over the course of a single person's lifetime than has happened throughout all of human evolutionary history.

But what happens to cancer evolution when we die? Can we really call what happens inside our bodies "evolution" if cancer cells ultimately kill the host they are living inside of? Can we say that a species evolved if eventually it went extinct? Of course the answer is yes: nobody would argue that dinosaurs didn't evolve because they eventually went extinct, or that a species that evolved itself into a dead end somehow nullified its previous evolution.

Just as species evolve into extinction, populations of cancer cells evolve in the body *until* they become evolutionary dead ends. In fact, cancer cells evolving themselves into evolutionary dead ends is an example of a more general phenomenon in evolutionary biology called "evolutionary suicide." In evolutionary suicide, populations of organisms can evolve traits that ultimately doom the entire species to extinction, like being so effective at consuming resources that they leave nothing left for future generations, or having such extravagant sexual ornaments that the entire population becomes catastrophically vulnerable to predation. And cancer cells are not always evolutionary dead ends either. Sometimes—as I will illustrate—cancer cells can transmit to other individuals and spread in the population. Transmissible cancers have been discovered in a number of species including domestic dogs, Tasmanian devils, and several species of clams. In all of these species, cancerous cells from one organism are able to leave the original host, transmit to a new host, and then grow inside that new host, making it possible for a cancer to survive much longer than the lifetime of the individual from which it derived and also making it possible for evolution to continue acting on the populations of cancer cells for much longer than it otherwise would. Nevertheless, transmissibility of cancer cells is not a prerequisite for evolution; the vast majority of cancer cells die with their hosts, and until that day comes, the cancer is subject to natural selection and drift, just like every evolving population is.

In order for a population of cancer cells to evolve through natural selection, certain conditions have to be met. These are the same conditions that have to be met for any population evolving in the natural world: variation, heritability, and differential fitness. In other words, cells must vary in their traits, these traits must be heritable when cells divide, and the traits must have effects on cell fitness (i.e., the survival and replication of the cells). Do cancer cells meet these conditions? They most certainly do. Cancer cells are a diverse population of cells with heritable traits that affect the fitness of the cells. Let's look a little closer at each of these conditions for natural selection and how cancer cells meet them.

Variation: A human being starts as a single cell. Many of us have been taught that as that cell divides, it copies its DNA, resulting in trillions of genetically identical cells that make up our bodies. But this isn't quite right. Our bodies are made of nearly identical cells, each having DNA that regulates proper multicellular behavior. But this DNA needs to get copied every time a cell divides, and the copying process isn't perfect. Every time a cell divides and copies its DNA, there is a chance that errors will occur, and some chance that those errors won't be found and corrected by our genetic "proofreading processes." Because of genetic mutations, cells in our body are not identical.

How different are these cells from one another? The variations are greater than you might think. Most cells in the body harbor unique mutations that result from DNA copying errors or other sources of mutation such as sun damage or chemical exposures. On top of this genetic variation, there is also epigenetic variation. Epigenetic variations refer to the differences among cells in terms of gene expression. In each of our cells some DNA is "exposed," making it possible for it to be read and translated into proteins, whereas other DNA is "bound up," staying silent and not producing proteins. These epigenetic differences also contribute to variation in how cells behave, for example, whether they move, consume resources, and send signals to their neighbors. Cells in our bodies vary both genetically and epigenetically, and this variation can serve as the fuel for somatic evolution.

Heritability: Heritability refers to a correlation between the traits of the parent and the traits of the offspring. If heritability did not exist, then any trait that helped biological parents survive and reproduce couldn't be passed on. Are genetic and epigenetic differences among cells heritable? Of course. Every time a cell divides, mutations in the DNA of that cell get copied and passed along to its cell progeny. Differences in DNA expression can be inherited as well, since epigenetic alterations to the DNA can also get copied and passed along when a cell divides. With these processes in mind, we can think of our bodies as a vast family tree of cells, in which the trunk of the tree is the first cell that gave rise to the fertilized egg, and each branch represents a cell division during which cellular traits are inherited (see figure 2.1). Mutations can be passed down in this cellular family tree just as they can be passed down from parent to child.

Differential fitness: Differential fitness refers to the idea that individuals with certain traits will produce more progeny than individuals with other traits. Do cells in our bodies differ with regard to how many cellular progeny they produce? Yes. From our earliest beginnings in the womb, some of our cells replicate more rapidly and extensively than others. Some tissues in the body exhibit more cell proliferation than others, and within the tissues, some cells proliferate more quickly than others. Many of these differences in proliferation are the result of normal epigenetic differences between cells (i.e., differences in gene expression) that allow us to properly develop into functional multicellular bodies with toes, ears, organs, and all of our other parts. Epigenetic differences among cells are necessary for us to develop normally, but they can also contribute to our susceptibility to cancer.

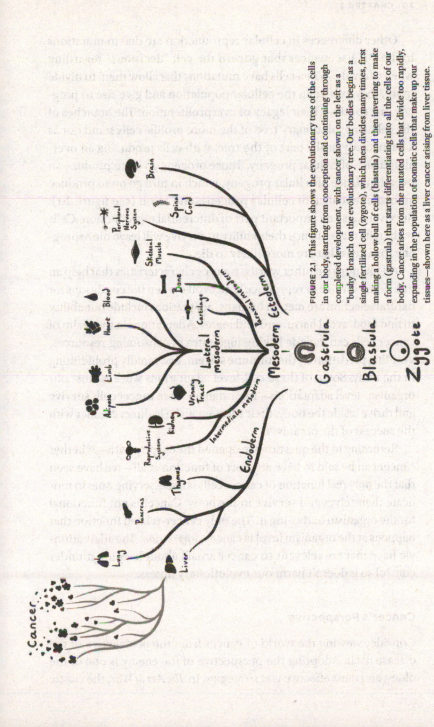

FIGURE 2.1 This figure shows the evolutionary tree of the cells in our body, starting from conception and continuing through completed development, with cancer shown on the left as a "bushy" branch on the evolutionary tree. Our bodies begin as a single fertilized cell (zygote), which then divides many times, first making a hollow ball of cells (blastula) and then inverting to make a form (gastrula) that starts differentiating into all the cells of our body. Cancer arises from the mutated cells that divide too rapidly, expanding in the population of somatic cells that make up our tissues—shown here as a liver cancer arising from liver tissue.

Other differences in cellular reproduction are due to mutations in the DNA sequences that govern the cell "decisions" regarding when to divide. When cells have mutations that allow them to divide more, they expand in the cellular population and give rise to progeny that continue their legacy of overproliferation. The branches of the cellular evolutionary tree of the more prolific cells stand out as a particularly bushy part of the tree, with cells producing an overabundance of cellular progeny. Those progeny go on to produce an overabundance of cellular progeny, which in turn go on to produce an overabundance of cellular progeny, and so on (see figure 2.1). Survival is also an important part of differential reproduction. Cells with traits that enhance their ability to survive will leave more progeny than cells that are more likely to die.

Adaptations (in other words, traits or characteristics that help an individual survive or reproduce) can evolve when the conditions for natural selection are met. In humans, adaptations include our ability to find food, avoid harm, and find mates. Adaptations in the realm of cancer cells can include having high rates of consuming resources, avoiding predation by the immune system, and rapidly proliferating in the body. Some of these cell-level adaptations work against our organism-level adaptations—for instance, when cancer cells survive and thrive inside the body, their success can be in direct conflict with the success of the organism.

Returning to the question we opened the chapter with—whether cancer can be said to have any sort of function at all—we have seen that the only real function of cancer cells is a self-serving one: to replicate themselves and survive in the body. Cancer is not functional for the organism harboring it. The only cancer-related function that happens at the organism level is cancer *suppression*. The adaptations we have that are relevant to cancer are all about keeping it under control so it doesn't harm our evolutionary fitness.

Cancer's Perspective

Consider viewing the world of cancer from the perspective of the disease itself. Adopting the perspective of the enemy is one of the oldest and most effective war strategies. In *The Art of War*, the classic work of military strategy, Sun Tzu warns against entering battle without knowing your enemy. In fact, if we know our enemy, we may discover that there is an opportunity for peaceful coexistence. Understanding both sides of a conflict can save you from engaging in protracted, unwinnable battles.

In the previous chapter, I argued that the war metaphor for cancer is problematic because it encourages a mentality of all-out destruction, which is often impossible in the case of cancer because populations of cells evolve resistance to our treatments. The idea that complete annihilation is the only way to win is simply a poor strategy. A better strategy is to look at a conflict from the perspective of the other side, to better understand its vulnerabilities, avoid costly escalation, and find a way to subdue the threat.

In our struggles with cancer, sometimes a decisive win is possible. Researchers and clinicians have achieved great success in the treatment of some childhood cancers and genetically homogeneous cancers in their early stages. But if a decisive win is not possible—because the cancer is too advanced, for example—and we attack cancer with massive force anyway, our bodies suffer the collateral damage of the toxicity of treatment. On top of that, an overly aggressive approach using high-dose therapies can lead us to lose control over the tumor because cells evolve to be resistant to our treatments. If we adopt the perspective of cancer, we can potentially guide ourselves to new and better strategies for controlling it.

So what does the world look like from the perspective of a cancer cell? From a cancer cell's point of view, our bodies are raw materials that can be consumed and used to make more copies of itself; from a cancer cell's point of view, our immune cells are predators to avoid, and our tissues and organs are new territories to be colonized. From a cancer cell's point of view, we are expendable. Cancer cells have no way to coordinate their behavior to keep from destroying us, their hosts. Cancer cells evolve anew inside every individual through a process of blind trial and error that eventually leads them to evolve adaptations (such as high proliferation rates and high metabolism) that can threaten the life of their host. As I've mentioned, cancer can evolve itself into an evolutionary dead end: cancer cells evolve to exploit their hosts, which can ultimately lead to their own demise.

As I discussed, cancer cells can evolve via natural selection inside the body, and natural selection can lead to adaptations. One way to think about adaptations is to consider this question: "How might this trait, characteristic, or behavior enhance the survival or reproduction of this entity?" This perspective is part of what is called the adaptationist approach in evolutionary biology. Adaptationism is a powerful tool for generating new hypotheses about how and why organisms evolved the way they did. We can apply the adaptationist approach to understand how cancer cells evolve in the body as well.

If we take the perspective of the organism, we can consider what would help it get its genes into the next generation. But this is simply a heuristic for reasoning about what an organism is doing and why. Organisms can be expected to act "as if" they have the intention to get their genes transmitted to the next generation because natural selection favors organisms that successfully achieve these goals. Likewise, we can think about cancer cells as if they have the goals of getting their genes to the next cellular generation. But we must keep in mind that this is just a mental tool to help us understand how cancer can evolve inside the body.

Thinking this way is called teleological thinking (*teleo* from the Greek word for "after," and *logos*, meaning "reason"). When we interpret the reason for an occurrence in terms of its *consequences*, we are adopting a teleological perspective. It is quite natural for us to think about things in terms of overarching purpose, and it often serves us well to look for reasons for an event after the fact. But teleological thinking can also lead us to conclusions that are downright wrong. Remember the game we played at the beginning of this chapter? You might have been tempted to choose one of the incorrect answers due to teleological thinking—inferring that the reason we get cancer has to do with the consequences that cancer has for us, for our kin, or for the human population. Inferring purpose where there is none is a pitfall of teleological thinking.

But teleological thinking is not always wrong. Something can happen because it serves some (not necessarily higher) purpose. Cancer cells, for example, have high rates of resource consumption and proliferation because these traits benefit cell-level fitness. In this sense, the purpose of cancer is simply to propagate itself.

An important part of adaptationist thinking in evolutionary biology, teleological thinking can be a useful heuristic for evolutionary biologists to develop hypotheses about organisms. An evolutionary biologist might look at the consequences of a trait and infer that it may have evolved to solve an adaptive problem. For example: Why do zebras have stripes? Perhaps to confuse their predators. Why do trees have leaves? To capture the energy of the sun via photosynthesis. If we understand the function of a trait that an organism has, we can better understand the evolutionary pressures that may have shaped it. But keep in mind that when a trait evolves because it enhances the fitness of the organism, this doesn't imply that any conscious intention was involved. The ancestors of zebras didn't *intend* to evolve stripes—it's just that the ones that had the more stripe-like pattern on their coats survived.

In the context of adaptationism, teleological thinking can be a useful starting point for developing hypotheses to test. But it can lead us to be overzealous in assigning evolutionary functionality to traits that may have no adaptive purpose. We should not assume that every trait and characteristic that a cancer cell has is an adaptation— some aspects of cancer are simply a result of random drift and do not involve natural selection or adaptation.

Overzealous teleological thinking can also lead us to assign evolutionary functionality at a level where it does not exist. Cancer doesn't exist to provide benefits for us as organisms or for the human population. Cancer does not evolve to enhance our fitness as humans or help the species survive—cancer cells exist only to survive and replicate. Taking a cell-level perspective can help us avoid the temptation of attributing reasons for cancer at the wrong level.

Understanding cancer from a cell-level perspective is a shortcut to an even smaller-scale perspective on evolution by natural selection: the gene-centric approach. This approach is fundamental in evolutionary biology, and was popularized by Richard Dawkins in his book *The Selfish Gene*, in which he described organisms as vehicles designed by natural selection to get genes into the next generation.

His basic argument is this: Genes that increase the survival or reproduction of the vehicles in which they reside will increase their numbers in the next generation. But this is not the whole story—natural selection will also favor "altruistic" genes that support the survival of other vehicles that share those same altruistic genes. In the case of human beings, natural selection can favor selfish individuals, as well as individuals who care for their kin. In the case of cancer cells, natural selection can favor selfish cells, as well as cells that provide benefits for their cellular brethren.

Both parts of this selfish gene approach—how it can shape selfish and cooperative vehicles—help reveal how cancer evolves in the body. The selfish gene approach is central to the argument of this book: that cancer is essentially a cellular cheater in multicellular cooperation. But the evolution of cooperation among cancer cells is just as important for understanding what happens during this progression. In later chapters we will see how natural selection can favor cancer cells that cooperate with one another to more effectively exploit their host, but first let's look more closely at how cancer evolves to cheat the cooperation that makes multicellular life possible.

3

Cheating in Multicellular Cooperation

There's nothing worse than sharing space with people who don't do their share, refuse to clean up after themselves, and won't listen when you try to talk to them about it. If you've ever had a bad roommate, you know how hard it is to coexist with somebody who doesn't contribute and cooperate fairly. If cancer cells in the body are like bad roommates, then cancer progression is like a nightmare roommate B movie. It starts with the roommate eating your food and not doing the dishes, letting the garbage and the dirty laundry pile up. Then it gets worse—much worse. One day you come home to find that the nightmare roommate has invited a lazy friend to stay over indefinitely, and the next day each of them has invited another lazy friend to stay, and so on in an exponential expansion of lazy roommates. They take over every room of the house, consuming everything in sight. Eventually there is no space left—but the horde keeps growing, trampling you as you struggle in vain to retain some semblance of control of the chaotic scene.

For normal cells living in a complex and well-regulated multicellular body, cancer cells are like a swarm of horrible roommates descending upon a perfectly regulated civilization and wreaking havoc. Cancer cells transform a multicellular body from a hardworking collective to a wasteland of exploitation, extortion, and conflict. As cancer cells divide, grow, and come to occupy more and more of the body, they can go from being the cellular equivalent of a bad roommate to being threats to the very fabric of the cellular society that makes us who we are as multicellular organisms.

But not all bad roommates destroy you. Some bad roommates are just clueless freeloaders, exploiting you out of their laziness or ignorance about the negative effects they are having. When the writer Jacob Brogan was diagnosed with thyroid cancer, he referred to his cancer as a quiet roommate who'd been living with him all along, occasionally leaving dirty dishes in the sink, but mostly staying out of his way. "We do not play host to cancer so much as we unknowingly sign a lease with it," Brogan wrote. Brogan's metaphor maps onto the biology well: often, we carry cancer around with us like a silent roommate, sometimes for decades, before it makes itself known.

This metaphor of a roommate—whether lazy, destructive, or mostly silent—is an alternative to the war metaphor for cancer, and one that is more appropriate for many cancers. A roommate who doesn't do their share is in many ways like a cancer cell that has stopped doing the jobs it should do as part of the multicellular body. A roommate who eats all your food is like a cancer cell that consumes the available resources. And a roommate who invites lazy friends to stay over is like a cancer cell proliferating out of control, burdening the body.

Like a freeloading roommate, cancer cells take advantage of the cooperative aspects of a multicellular body. And like a difficult roommate, cancer is not necessarily something to destroy. Rather than aiming for annihilation, we can in many cases set our eyes on a peaceful, if uneasy, coexistence. We can learn to live with cancer more effectively, and even shape it into a disease that is easier to live with.

In this chapter I will look at cancer as the bad roommate in the body—a cheater, sometimes taking advantage of the efforts of the normal cells in the multicellular body and other times actively exploiting them. Cheaters break the rules, both implicit and explicit, for how to behave when sharing space with someone. Likewise, cancer cells are cheaters because they break the rules of cellular sociality that make multicellularity possible. For the purposes of this chapter, and the book more broadly, I define cheating as breaking shared rules for the fitness benefit of the rule-breaker.

When I talk about cancer as a cheater, I'm not implying that cancer cells are intentionally breaking rules—only that they evolve to break rules because it is evolutionarily advantageous to do so. As I discussed in the last chapter, natural selection can shape populations—whether of organisms in the natural world or cancer cells in the body—so that individuals behave *as if* they have goals and intentions. Talking about cancer cells as cheaters is shorthand for saying that they evolve to behave in an exploitative way, breaking the rules of multicellular cooperation and getting a benefit for themselves at the expense of the body.

With the roots of our analogy in place, I'll discuss how cancer is a free rider in the cellular cooperation that makes multicellularity possible, and how evolution in the body favors cheating cells. First let's take a look at how cancer has been defined in the past and how the framework of cellular cheating fits in with different definitions and approaches to cancer.

What Is Cancer?

Cancer is as elusive to define as it is to treat. Sometimes cancer is defined as invasive growths, other times physicians use the term "cancer" for noninvasive growths. Sometimes doctors consider disrupted tissue architecture as the defining characteristic; other times they focus on certain key genetic mutations. My perspective on cancer is founded in the evolution of multicellularity and how cancer evolves as a "cheater" in the context of multicellular cooperation. This definition of cancer as a cheater provides us with an organizing framework for the many diverse perspectives, definitions, and approaches to cancer, spanning those grounded in genetics, cell biology, and comparative biology.

Definitions of cancer vary significantly—a cancer biologist, a pathologist, a clinician, and a comparative oncologist will all explain cancer in different ways, focusing on different elements of the disease. A cancer biologist may focus on the traits and characteristics of cancer cells, considering whether the cells have gained the ability to replicate indefinitely, produce their own growth factors, and avoid apoptosis (controlled cell death). Cancer biologists see these traits and several others as the hallmarks of cancer. The hallmarks of cancer were chosen because they appear over and over again across different kinds of cancers, and they now comprise an authoritative list of cancer traits (see box 3.1). I will explore how these hallmarks of cancer map onto the cheating behavior of cancer cells in several aspects of multicellular cooperation (figure 3.1).

But the hallmarks approach is not the only way of looking at cancer. If you ask a pathologist to define cancer, they are likely to tell you that cancer is defined by how the cells look under a microscope, and, in particular, whether the tissue architecture is abnormal and the cells are not properly differentiated.

As someone who treats patients directly, a clinician is likely to tell you that the defining feature of cancer is its invasive nature, and its ability to metastasize, because this is what is most important when providing a prognosis. If you ask a comparative oncologist— someone who studies cancer across species—they might first lament the challenges of defining cancer across species because different organisms have different underlying tissue biology. With that in mind, they may tell you, it doesn't make sense to use invasion and metastasis as criteria for cancer, because there are many species with cancer-like forms that have tissue architectures that make invasion and metastasis unlikely or impossible.

These varied definitions and approaches don't really tell us anything about cancer's fundamental nature or how it got to be that way. They tell us what cancer looks like, how cancer cells behave—but not what cancer really is. This is where the perspective of cellular cheating comes into play.

Seeing cancer as a cellular cheater helps unify the diverse perspectives on the disease and shows how cancer's nature relates to fundamental tensions in the evolution of cooperation and the evolution of multicellular life. The cellular cheating viewpoint can

BOX 3.1 The hallmarks of cancer and cellular cheating

The hallmarks of cancer were laid out in a landmark paper by the cancer biologists Douglas Hanahan and Robert Weinberg in 2000. They identified six hallmarks, which they updated a decade later to include two emerging hallmarks and two enabling characteristics. These hallmarks include characteristics like resisting cell death, enabling replicative immortality (dividing indefinitely), altering cell energetics (e.g., increasing resource use), and enabling metastasis by breaking down the normal tissue architecture. These hallmarks of cancer also map onto cellular cheating in the foundations of multicellular cooperation (figure 3.1), or breaking of the rules in the "playbook of multicellular cooperation."

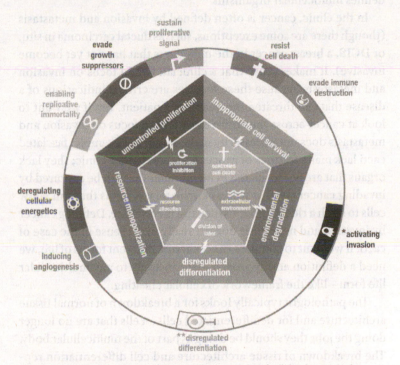

FIGURE 3.1 The hallmarks of cancer fit into the five categories of cellular cheating in the foundations of multicellular cooperation. The hallmarks are shown on the outer circle, and the foundations of multicellular cooperation are shown on the inner pentagon. The types of cheating in multicellular cooperation—uncontrolled proliferation, inappropriate cell survival, resource monopolization, dysregulated differentiation, and environmental degradation—are shown in the outer pentagon. Each of these aspects of cellular cheating corresponds to characteristics of cancer. The hallmark of invasion is shown with an asterisk because it typically involves cheating in multiple foundations of cellular cooperation. Dysregulated differentiation is shown with a double asterisk because it is not currently named as a hallmark of cancer, but the cellular cheating framework suggests we should perhaps consider it a missing hallmark (dysregulated differentiation is considered an important characteristic of cancer by pathologists, also suggesting that it should be added to the list of cancer hallmarks). Figure reprinted with permission (Aktipis 2015, licensed by CC BY 4.0).

bring multiple perspectives together into one framework. As we saw earlier, cancer biologists often look to the hallmarks of cancer, a list of cancer cell phenotypes (traits and characteristics), as an authoritative description of the disease. These cancer hallmarks are manifestations of cheating in the cellular cooperation that usually defines multicellular organisms.

In the clinic, cancer is often defined by invasion and metastasis (though there are some exceptions, such as ductal carcinoma in situ, or DCIS, a breast cancer in the milk ducts that has not yet become invasive). It makes sense that a clinician would focus on invasion and metastasis because these features are critical indications of a disease that can threaten the health of a patient. But if we want to look at cancer across the entire tree of life, a focus on invasion and metastasis does not necessarily make sense. For example, fasciated cacti lack many features of cancer as defined in the clinic; they lack organs that are encapsulated in membranes that can be ruptured by invading cancer cells, as well as circulatory systems that can allow cells to hitch a ride to other tissues and metastasize. Defining cancer by invasion and metastasis doesn't really make sense in the case of cacti. If we want to compare cancer across different forms of life, we need a definition and approach that can apply to any multicellular life form—like the framework of cellular cheating.

The pathologist typically looks for a breakdown of normal tissue architecture and for dedifferentiated cells—cells that are no longer doing the jobs they should be doing as part of the multicellular body. The breakdown of tissue architecture and cell differentiation represent breakdowns in two important aspects of multicellular cooperation: taking care of the extracellular environment and dividing labor. Dedifferentiation, in particular, can influence somatic evolution because cells that don't differentiate can continue replicating, whereas cells that differentiate typically replicate a few times and then become part of normal tissue (until they die and slough off).

By focusing on the ways that cellular cooperation can break down, we can come up with an overarching framework that spans multiple fields and allows us to compare and contrast cancer susceptibility across the entire tree of life.

The Evolutionary Puzzle of Cooperation

How can cooperation evolve through natural selection if cheaters receive higher payoffs than cooperators? To understand the evolutionary reasons why cancer is a cheater in multicellular cooperation, we have to look at the broader question of how cooperation can evolve and remain stable in general.

Cooperation theorists have proposed many solutions—from reciprocity to punishment and risk-pooling to cooperating only with kin—and they have explored potential explanations through hundreds of computer models. Despite the diversity of the solutions and strategies for making cooperation stable, they mostly fall into two main categories: solutions based on individuals having repeated interactions with one another, and solutions based on genetic relatedness.

When individuals repeatedly interact with one another, this can favor cooperation because there is an opportunity to reap the benefits of past cooperation or experience the negative repercussions of having cheated. Repeated interactions change the payoffs for cooperation and cheating—often making cooperation a better option overall. When individuals can leave uncooperative partners and groups—or engage in any sort of partner choice—the benefits of a cooperative strategy can increase. As a strategy, cooperation can be better than cheating because groups of cooperators are more stable and longer-lasting, reaping the benefits of cooperation. When it comes to understanding the evolution of multicellularity, repeated interactions of cooperative cells may have played a role in the early stages of the evolution of multicellular life. But traditionally, scholars see genetic relatedness as the main explanation for the evolution of cell cooperation in the transition to multicellularity.

Genetic relatedness helps solve the problem of cheating, and makes cooperation viable by allowing its benefits to come back to the genes coding for it. Imagine a soup of unicellular cells (cells that are not part of a multicellular organism). Some of these cells are "producers," meaning that they produce something that enhances the fitness of every other cell around them (such as an enzyme that helps the cells metabolize resources). Other cells are "free riders" (or "freeloaders"; free rider is the term that cooperation theorists typically use to denote an individual who is not contributing to a public good), meaning they do not produce anything, but benefit from what the producers make. If interactions happen at random among individuals in the population, free riders will get more resources and avoid paying the costs of producing the enzyme. They will have more opportunities to reproduce (because they have more resources) and will come to dominate the population. Eventually, free rider cells will take over and all the cooperative cells will go extinct. This is a prime example of the classic problem of cheating and how it limits the evolutionary viability of cooperation.

But consider an alternative to random interactions among the cells in this primordial soup: What if producers stick together and interact with one another and not with the free riders? Whenever the producers make something, the benefit goes to other producers. Rather than feeding free riders, producers provide benefits for their own kind. This allows the genes that code for the ability to produce to increase in the population of cells.

Similarly, when all the cells in a cellular group are genetic clones, genes that code for cellular cooperation can spread through a process called kin selection. One of the reasons that extremely high levels of cellular cooperation evolved in multicellular bodies is that the cells are (at a first approximation) genetic clones. Genetic relatedness is not the whole solution—as we will see in the next section—but it does help to set up conditions that make the evolution of cellular cooperation possible. Having high relatedness inside a cellular group also makes it possible to evolve mechanisms for detecting and dealing with cellular free riders.

Genetic relatedness of clusters of cells made it more likely that cellular cooperation could evolve, setting the stage for the evolution of multicellularity. But what made multicellularity a good strategy in the first place? Why would cells ever give up their ability to reproduce themselves as individuals and subordinate their cell-level evolutionary fitness to that of the cellular collective?

Multicellularity Is Cooperation Incarnate

Have you ever considered how much easier life would be if you could clone yourself? One of you could go to work, one could stay home and take care of the dishes and the laundry, and a third could catch up on email. And for that matter, why stop at three? Why not create a whole army of clones to get everything done?

This is why life on Earth made the jump from unicellular living to the multicellular lifestyle: it made life easier. Early in the history of this planet, unicellular life such as algae and bacteria dominated, replicating and using resources like carbon and nitrogen, going about their unicellular lives. But some cells tried a new strategy: when they divided, they stuck together instead of separating into two separate cells. Eventually these clumps of cells evolved the ability to divide labor through regulating the genomes of the cells inside: some cells specialized in moving the organism around, some specialized in digesting food, others specialized in reproduction. This division of labor allowed multicellular life to be much more efficient than life as a single cell.

This is of course an oversimplification of the evolution of multicellularity. Early multicellular groupings of cells had many other advantages over solitary cells (for example, the ability to avoid predation and manage risks by sharing and storing resources). Groups of cells that could coordinate themselves as a collective could survive and thrive. Cooperation in the form of multicellularity was an effective strategy, and so multicellular life thrived and expanded into many ecological niches on our planet—from the deepest seas to the highest mountains and everywhere in between.

The initial evolution of multicellularity opened the door to the eventual evolution of large and complex multicellular life such as human life. With a huge society of cellular clones working toward the common goal of keeping the multicellular organism alive and well (and reproductively successful), organisms could achieve massive division of cellular labor, movement on an unprecedented scale, and the evolution of complex nervous systems that can rapidly process and respond to information—just as you are doing now as you read this book.

But for all of these benefits to multicellularity, there are also problems. Big problems. The larger a cellular society is, the larger a target it is for would-be exploiters—specifically, cells that can benefit from cheating. This classic type of cheating plagues cooperative systems, but there are several solutions to manage the problem. One of them is genetic relatedness. If an individual shares genes that code for cooperation (for example, genes for producing a public good) with a biological relative, then copies of those genes in relatives get the benefits of cooperation, leading these genes for cooperation to increase in the next generation. Multicellular bodies solve this problem of cellular cheating in part through genetic relatedness. At first approximation, our multicellular bodies are made of genetically identical cells that come from a single fertilized egg, and so genes coding for cellular cooperation and control of cellular cheating can persist.

But genetic relatedness is not enough to ensure effective cooperation and coordination. Imagine an army of clones of yourself for a moment: Which of you would be in charge? Would the rest of your clones obey? How would you coordinate and assign tasks, or share information to effectively accomplish goals? What would happen if one clone was dishonest or unkind or just plain lazy? Presumably you and all of your clones would share the same goals and interests—as the cells in our body do—but sharing the same goals doesn't necessarily solve the problem of how to organize and coordinate your activities. In addition, if there was some variation in skill and motivation among your clones, it would be difficult to figure out who (if anyone) was free riding (and if so, what to do about it). These problems of coordinating, regulating, and monitoring a society of clones are the same problems that multicellularity had to solve in order to become large, long-lived, and complex.

Cells in multicellular bodies regulate and coordinate their behavior through complex signaling systems and genetic networks that keep them from harming the collective. Because all cells in the body share (mostly) the same DNA, they share the same systems for regulating and coordinating cellular behavior. These systems can be thought of as a playbook for multicellular living. This playbook does not dictate what every cell will do at every moment, but it provides a menu and some guidance for how to respond to different situations.

Effective multicellular cooperation is based on some foundational cell-level behaviors that enable the body to develop and function (figure 3.1). My colleagues and I have called these the "foundations of multicellular cooperation" in previous publications, but I prefer the playbook metaphor for our discussion here because it clarifies that these cooperative characteristics are actually behaviors that make multicellularity possible.

What is in this multicellularity playbook?

1. **Don't divide out of control.** In order to develop as a coherent and functional multicellular body, cells have to inhibit proliferation/division. Without this control on cell proliferation, multicellular structure and function would be compromised and multicellular organisms would continue to grow indefinitely.

2. **Self-destruct if you're a threat.** Some cells can threaten the viability of a multicellular body. Mutated cells that are dividing out of control are one such example. Other cells, such as those that initially form webs between our fingers and toes during development in the womb, may be in the way. Self-destruction—in the form of controlled cell death—allows these cells to quietly eliminate themselves.

3. **Share and transport resources.** In multicellular organisms that are more than a few millimeters across, oxygen and other nutrients can't reach cells on the inside simply through diffusion; some sort of active transport of resources is required. For example, our digestive systems and circulatory systems are complex resource transportation systems that allow cells in our bodies to access the nutrients they need to survive and do all the jobs that are necessary to make us viable multicellular organisms.

4. **Do your job.** One of the foundations of multicellular cooperation is the division of labor. There are hundreds of

cell types in our bodies, each of which performs a different job: liver cells detoxify the blood, heart cells pump blood, neurons transmit electrical signals. Sometimes cells stop working—or stop doing their jobs correctly. These cells can threaten the multicellular body since they can express the wrong genes at the wrong times and wreak havoc on the broader regulatory systems that make multicellularity work.

5. **Take care of the environment.** Our bodies are worlds unto themselves. Our cells create a tissue architecture that they live within, and they have systems for gathering up and eliminating waste that would otherwise accumulate in the body. Cells in our bodies create those internal worlds during our development and then maintain them throughout our lives, keeping the architecture in place and removing waste. Tissue architecture helps keep cells where they should be (preventing them from invading neighboring tissues) and keeps cells in the right expression state so that they produce the right proteins and do the right jobs.

These five basic rules in the multicellularity playbook are fundamental to the life and health of multicellular organisms. When they are disrupted, the stage is set for cancer. So what does the breakdown of this multicellular cooperation look like?

Sometimes the genetic machinery underlying the multicellularity playbook can get damaged. This damage can occur as a result of genetic changes like DNA mutations or through epigenetic changes (such as abnormal gene expression). Damaged cells that aren't following the rules of multicellular engagement can sometimes get evolutionary benefits from exploiting the cells that *are* following the rules of multicellular cooperation. It's important to note that damaged cells are rarely evolutionarily advantaged. Usually mutations make the cells less viable, and, even if they confer a benefit (say in the form of increased rates of proliferation), these mutations often make them targets for destruction. Our body has systems that detect and eliminate cells that are potential cancer risks, and this usually eliminates any potential advantage a mutated cell might have. Nevertheless, damaged cells occasionally gain an evolutionary advantage over normal cells. Let's take a look at several examples.

Controlling cell proliferation is an essential component of multicellular cooperation—it keeps the multicellular body coherent, stable, and hopefully cancer-free. Rapid cell proliferation is one of the central hallmarks of cancer. In chronic myeloid leukemia there is often a mutation called a translocation, which "rewrites" this playbook of multicellular behavior, taking one chunk of chromosome and putting it next to another chunk of chromosome where it doesn't belong. This translocation results in a fusion gene called *BCR-ABL*, in which the *BCR* gene's promoter (the part of a gene that tells the gene to turn on) joins with the *ABL* gene's proliferative signal (which is responsible for cell proliferation in the immune system). As a result of this fusion, the cell reads the gene sequence as an instruction to keep proliferating. Cells possessing this mutation keep proliferating when normal cells would not, and because of this mutation, they don't follow the same rules that other cells do. As a result of this cheating, they leave more cellular descendants behind.

Cancer can also be a downstream consequence of mutations that damage genes regulating cell death, like the gene *TP53*. *TP53* is cancer suppressor gene I will use as an example throughout this book. It helps protect the multicellular organism from damaged cells by causing cell death if the cell's DNA is mutated beyond repair. But if *TP53* itself is mutated, then cells can keep surviving and proliferating even if they have severely damaged DNA. Thus, damage to *TP53* and other genes that regulate cell death can confer an evolutionary advantage to the cells harboring that damage. In the end, cells that can cheat cellular death have an advantage over cells that follow the rules of the playbook, which tell them to die if they threaten the viability of the organism.

Disruption of proliferation inhibition and controlled cell death are just two examples of how the playbook of multicellularity can get damaged and how this damage can lead to cancer. Damage to other aspects of the multicellular cooperation playbook—for example, the genes encoding the rules of multicellularity that regulate resource use, division of labor, and the maintenance of the extracellular environment—can contribute to cancer risk as well. Cancer cells usually have mutations in genes that regulate resource use—in the form of mutations in metabolic pathways—allowing cells with the mutations to consume much more than the normal cells that are following the rules. Disruptions to the cellular division of labor can also contribute to cancer: if cells do not differentiate into the appropriate cell types, or if they dedifferentiate (in other words, revert to a stem-like state that can turn into any kind of cell), this disruption to normal differentiation can wreak havoc on the body, altering tissue architecture and compromising organ function. In addition, cells that aren't doing their jobs are expending less energy, which means they have extra resources that they can put toward proliferation or other ends that threaten the viability of the multicellular body. Finally, cancer cells can neglect and even actively destroy the internal environment of the body through, for example, producing lactic acid. Lactic acid can break down the extracellular matrix, destroy the tissue architecture, and even allow cancer cells to invade neighboring tissues.

One of the reasons why cancer evolution is so complicated is that natural selection happens on two different spatial levels and timescales: among cells inside organisms over the relatively short timescale of the organism's lifetime, and among organisms over the very long time scale of organismal evolution. Cancer cells evolve in the body, but bodies that are better at suppressing cancer—through having systems that effectively detect and eliminate cellular cheating—survive better and leave more descendants. Understanding multilevel selection—evolution happening on different levels of organization, like cells and bodies—is essential to understanding the puzzling aspects of cancer.

Cancer is a process of cheating cells evolving inside the body, but cancer suppression also evolves because organisms that are better able to keep cellular cheating under control typically survive longer and therefore have more opportunities to reproduce. When natural selection acts on multiple levels of organization, we refer to the process as multilevel selection. Let's look at the classic case of a social dilemma to understand what happens to cooperation and cheating in the case of multilevel selection. A social dilemma is a situation in which the strategy that is optimal for the individual is different from the strategy that is optimal from the perspective of the group. We see a similar phenomenon in the case of cancer: what is optimal from a cell-level perspective (cheating in multicellular cooperation) is different from what is optimal from the organism-level perspective (following the playbook of multicellular cooperation).

Imagine a population of individuals, including some cooperators and some cheaters. Let's say there are 100 individuals in 10 groups with 10 members each. This kind of structure is typically called a metapopulation. Initially, the population starts with about half cooperators and half cheaters overall, distributed randomly among these 10 groups (figure 3.2). Due to random variation, some groups will include a majority of cooperators, others will have mostly cheaters, and some will be a mixture of cooperators and cheaters. Every group is playing a public goods game—each individual can choose to invest in the good of the group at an individual cost (what cooperators do) or choose not to invest (what cheaters do). No matter how much each individual puts into the group, each gets the same-size proportion of the group good. In public goods games like this, cooperators invest in the good of the group, and cheaters free-ride on the cooperation of their group members.

Now let's add evolution via natural selection to this scenario: Individuals who get higher payoffs are more likely to survive and reproduce, creating copies of themselves in the next generation (see figure 3.2). This means that within each group, cheaters will outperform cooperators, and cheaters will proliferate. This process is very similar to what happens in cancer.

But something quite interesting and paradoxical emerges if we look between groups. Within each group, cheaters have a clear advantage over cooperators, slowly but surely taking over the group they are in. But if we look at the population as a whole, groups that have more cooperators grow, whereas the groups that have more cheaters shrink. Therefore, cooperators can win in the population as a whole, even though they have a disadvantage compared to cheaters within any given group. If these groups can divide and replicate themselves, the groups that have fewer cheaters end up leaving more descendants.

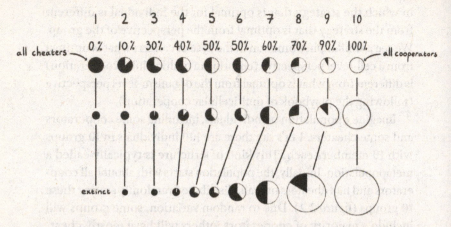

FIGURE 3.2 Cooperators can outperform cheaters when the population is subdivided into several groups. The initial composition of groups is shown at the top, with changes in the group size and composition shown going down from the top. Groups that begin with all or mostly cheaters (*left*) shrink over time, while groups that begin with all or mostly cooperators (*right*) grow over time. Looking across the whole population, cooperation increases over time, even though within each group cheaters increase in frequency. This apparent paradox is resolved by the fact that groups with more cooperators grow more quickly than groups with more defectors.

A population of multicellular bodies is essentially a metapopulation of cells that is partitioned into groups of approximately thirty trillion cells. Within each multicellular body, selection favors cells that cheat. But in a population of multicellular bodies, the bodies with more cellular cooperators win: they survive longer and leave more offspring. What we end up with is selection favoring organisms that are made of cells that are better multicellular cooperators, and better at detecting and suppressing cellular cheating within. Natural selection at the organism level favors cellular cooperation—and the ability to detect and suppress cellular cheating. In other words, selection favors organisms with better cancer suppression systems.

Not only does selection operate at two very different levels—favoring organisms made of cellular cooperators and favoring individual cells that cheat—it is happening at two very different time scales. Organisms have been under selection for hundreds of millions of years to be effective cellular cooperators and be good at suppressing cancer. But cells are under selection over the course of our very lifetimes, with cheaters having an evolutionary advantage over cooperators within our bodies.

This process of multilevel selection (also called group selection) is an accepted fact, as applied to cancer cells and multicellular bodies. But I would be remiss not to mention that there is still controversy about whether this kind of process has shaped human populations—specifically whether cooperative groups of humans were favored over groups of cheaters in human evolutionary history. What is most relevant to the ideas here, however, is that there is no debate about the importance of multilevel selection for understanding how cancer cells evolve and how multicellular bodies evolve to suppress and control cancer itself.

Finding Cellular Cheaters

Cancer cells will always have an evolutionary leg up on organisms because cells can replicate much more quickly than organisms can: cells divide every few days, but multicellular organisms like us reproduce every few decades. With this in mind, cancer cells have an advantage when it comes to the sheer speed of evolutionary change. But we organisms have an advantage when it comes to the complexity of our strategies for controlling cancer. Our multicellular bodies have been evolving for millions of years to anticipate cancer evolution, so we have some tricks up our sleeves to keep would-be cancer cells in check. Evolution has equipped us with an arsenal of cheater detection mechanisms that our bodies use to suppress cancer. From DNA repair mechanisms to cell division control systems and immune surveillance, our bodies have evolved redundant enforcement mechanisms to keep misbehaving cells at bay.

CELLULAR CONSCIENCE

Cells have intrinsic mechanisms that keep their behavior in check—the cellular equivalent of a conscience, monitoring the internal state of the cell and letting the cell know if it becomes a risk to the body as a whole. These mechanisms monitor the cell's internal state for any abnormal behavior that might indicate the cell is a threat to the integrity or viability of the multicellular organism. They allow a cell to constantly monitor itself to make sure it behaves properly as a member of the multicellular body. Of course, this monitoring is not happening in a conscious way—the cell is simply processing information through its genetic network, a sophisticated information processing system that allows the cell to take these kinds of abnormal behaviors as inputs and then output an alarm signal to other parts of the genetic network if there is a problem.

This information is routed through genetic networks like that of the cancer suppressor gene *TP53*. Cancer suppressor genes like *TP53* and the networks of information that feed into them are designed to detect DNA damage, aberrant proteins, or other signals that could indicate that a cell has gone awry and is no longer serving the fitness interests of the multicellular body. *TP53* is a central node in a vast network of cellular information, acting as a central intelligence agency to monitor the body's cells (figure 3.3). It integrates signals from all over the cell and the neighborhood to "decide" what each individual cell's fate should be. Cancer researchers have called *TP53* "the guardian of the genome," though "the cheater detector of the genome" is my preferred way to think about it. *TP53*, if activated, can block the cell from replicating, jump-start DNA repair, and, if the cell is too damaged, initiate apoptosis (in other words, programmed cell death).

I will come back to *TP53* when I discuss cancer across the tree of life. Differences in cancer suppressor genes like *TP53* play an important role in susceptibility to cancer among species. Elephants, for example, have multiple copies of *TP53*, and this is likely one of the reasons that they are particularly resistant to cancer. Unfortunately, cell intrinsic cancer systems such as *TP53* can fail—in all organisms from elephants to mice to humans—and cells with DNA damage can continue to survive and proliferate. If this happens, our bodies need to move to the next line of defense: neighborhood monitoring.

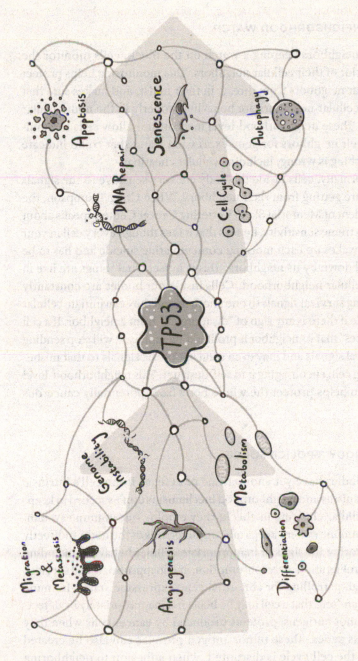

FIGURE 3.3 The cancer suppressor gene *TP53* is a central node in the genetic network that processes information to "decide" whether a cell is a threat to the viability of the organism. *TP53* essentially functions as a cheater detector system inside every cell. Through production of the protein p53, it is able to gather information from many different aspects of cell function that could indicate cellular cheating (including abnormal metabolism, genomic instability, and inappropriate migration). In response to this information, *TP53* can halt the cell cycle, repair DNA, and even induce apoptosis (cellular suicide) if necessary.

THE NEIGHBORHOOD WATCH

Like neighbors keeping a watch on the block, cells monitor the behavior of their cellular neighbors. This monitoring helps protect cellular neighbors from threats in their midst, and makes sure that their cellular neighbors are behaving properly in the multicellular body. These neighborhood-level mechanisms allow cells to monitor their neighbors for gene expression states that could indicate something is wrong, including cellular cheating.

Normally, cells inside the body are very sensitive to the signals they are getting from their neighbors. When Craig Thompson, the president of Memorial Sloan Kettering Cancer Center, speaks about this extreme sensitivity, he says that it is as though every cell in your body wakes up each morning contemplating suicide and has to be talked down by its neighbors. This is in fact what things are like in the cellular neighborhood: Cells inside our bodies are constantly sending survival signals to one another, and they can initiate cellular suicide if there is any sign of "disapproval" from a neighbor. If a cell "notices" that its neighbor is proliferating rapidly, it will stop sending survival signals and may even send apoptotic signals to that misbehaving cell, encouraging it to self-destruct. This neighborhood-level system helps protect the whole body from potentially cancerous cells.

THE BODY'S POLICE FORCE

Our bodies have yet another line of defense if the cell's intrinsic mechanisms and neighborhood mechanisms don't succeed in keeping cellular cheating in check: they employ our immune system. The immune system keeps tabs on all regions of the body, indirectly monitoring for signs of inappropriate cellular behavior, including overproliferation, overconsumption, and inappropriate cell survival, through patrolling for abnormal gene expression states. Immune cells can "see" that a cell may be behaving inappropriately by detecting tumor antigens, proteins produced by cancer cells when they express genes. These tumor antigen proteins can also be created when the cell cycle is disrupted, when adhesion to neighboring cells is disrupted, and during cellular stress responses. The immune system collects information about cellular behavior across all the tissues and organ systems, looking for signs that something might be wrong—like the presence of these tumor antigens—and mobilizing immune cells to any location where there might be a problem. Immune cells are on a search-and-destroy mission for anything that is damaging to the multicellular body, and cancer cells are no exception. If the immune system can find and identify cancer cells, it can target them for elimination and help protect the body from the threat of cancer.

Ideally, these three cancer suppression systems—intrinsic, neighborhood, and systemic—function together to detect and control would-be cancer cells. These mechanisms for detecting and suppressing cellular cheating usually work, but they are not fail-safe. These mechanisms are encoded by genes that can themselves get disrupted during cancer progression. And many of these cancer suppressor genes are mutated even in "normal" tissue with no sign of cancer-like growths.

As a result of mutations, populations of cancer cells can evolve to avoid all of these systems. For example, the cancer suppressor gene *TP53* can become mutated, leading to a failure of a cell's intrinsic cancer suppression systems. Mutations can happen in the genes that code for intercellular communication, leading cells to stop paying attention to signals from their neighbors. And the ability of the immune system to find cancer cells is constantly being eroded during cancer evolution. Cancer cells evolve to evade the immune system by changing the proteins on the outside of the cell or disrupting immune signaling. Like populations of animals that are constantly evolving to be better at escaping their predators, populations of cancer cells are under constant selective pressure to evolve ways of evading our defenses.

The Cellular Intelligence Agency

Cellular cheater detection systems work together to find, suppress, and eliminate would-be cancer cells, keeping us healthy and cancer-free. We have vast information processing networks in the form of cell signaling systems that we can use to police our multicellular bodies. Information is processed and transferred among cells, but also *within* cells in genetic networks like those around *TP53*. Large organisms such as humans have found a way to leverage cellular and genetic computation to detect and respond to cell-level cheating.

This information processing approach helps make sense of some of the evolutionary puzzles of *TP53*, in particular why it is both so important to cancer suppression and so vulnerable to being disrupted. *TP53* is perplexing: it plays such an important role in protecting the body from cancer, yet we have only one copy from each parent—if a copy is missing or mutated, the body is very vulnerable to cancer (as happens in Li-Fraumeni syndrome, a disease I will explore later in this book). *TP53* is also a central node in the genetic network of the cell, so if it is damaged, the consequences can be disastrous for regulating cell division and apoptosis. Wouldn't it make sense to have a more robust and redundant detection system instead? Why route so much information through *TP53*? Wouldn't it be logical to have a distributed system?

A potential solution to this conundrum comes from signal detection theory and what is referred to by evolutionary medicine researcher Randolph Nesse as the smoke detector principle—the idea that it makes sense to calibrate an alarm so that it is sensitive enough to detect a fire, even if that means tolerating some false alarms. In the case of cancer suppression, a false alarm that leads you to lose one "good" cell out of millions is a small price to pay for overall protection from cancer.

Of course, false fire alarms mean being awakened by blaring sirens, missing hours of sleep, or experiencing the annoyance of having to evacuate a building. The same is the case for false alarms from *TP53*; they come with some costs as well. If *TP53* "sounds the alarm" when there is no potential cancer threat from the cell, it can lead to premature aging. It can also cause excessive inflammation in the regions where apoptosis is happening, or even prompt the emergence of new, potentially cancer-causing mutations as cells proliferate to replace those that have apoptosed.

So, what is an organism to do? How should the organism "decide" between letting potential cancer cells survive versus killing some healthy cells and damaging the organism? Signal detection theory states that there is a fundamental trade-off between these two types of errors—misses (letting a cancer cell live) and false alarms (unnecessarily killing a healthy cell). But there is also a way to get around this trade-off: increasing accuracy with more effective use of intelligence (in other words, using more and better information to make the decision). In signal detection theory—a general framework for looking at decision-making in situations with ambiguous information—it is possible to improve accuracy by looking at multiple pieces of information, called cues, and combining them to make a better decision than could be made with just a single piece of information. In the case of cancer and *TP53*, it is possible to make "better decisions" by collecting information from many components of the genetic network that can help determine whether a cell is likely to be a cancer threat.

By integrating information from several sources, signal detection systems can be more accurate, reducing the likelihood of both false alarms and misses. The simplest way to do this is to integrate information from two sources rather than just having a detection threshold for one criterion. For example, fire alarms that integrate information about both smoke and heat can save lives by making it easier to detect the fire (reducing misses), and decrease the number of nuisance alarms (reducing false alarms). Integrating multiple criteria makes it easier to detect the signal (in this case, the fire) without setting off lots of false alarms as well. This kind of multicriterion detection can help increase the accuracy of any decision-making system. When the costs of false alarms are high, using more information to make a decision can help avoid the costs of overreacting. In the case of a fire alarm, overreaction is simply the alarm going off when there is no actual fire. In the case of our cancer suppression systems, overreacting would mean killing cells that are not cancerous. This kind of overreaction of cancer suppression systems like apoptosis may contribute to premature aging (and lead to potential trade-offs between cancer risk and aging). Using more information can help the cells (and the body as a whole) make better decisions about whether a cell is likely to be a threat.

Using information from multiple sources also means that one must have an effective way to integrate and weigh each criterion in order to make a decision. In the case of our multicriterion smoke detector, the system has to specify how to integrate information about smoke and heat: Should it go off only when both are above a certain threshold? Can just smoke or just heat set it off if the levels are high enough? Even for a two-criterion decision system like these advanced smoke alarms, the rules for when to set off the alarm become much more complicated, and the information has to be centralized and integrated in order to get the benefit of a more accurate decision.

More accurate decisions require more computational power. The processing system has to bring together the information in a circuit made of wires, neurons, or genes in a genetic network in which the criteria can be properly weighted, integrated, and evaluated. In the genetic networks that operate in our cells to help identify and eliminate cellular cheating, multiple criteria can be brought together to help distinguish between normal cells in unusual conditions and potentially cancerous cells.

Let's look at a situation in which a cancer suppressor gene that integrates multiple pieces of information can do better than a cancer suppressor gene using a single piece of information. If a cancer suppressor gene only has access to the information about how quickly a cell is dividing, and the threshold is set quite low so it doesn't miss any potential cancer cells, it might "sound the alarm" when a cell is proliferating rapidly while that cell is in the process of healing a wound. And if the threshold were higher, it might avoid this kind of false alarm, but miss a rapidly proliferating cell that actually poses a threat. A genetic network using just one criterion faces this fundamental trade-off between misses and false alarms.

But now consider a network that decides if a cell poses a cancer threat using two criteria: the proliferation rate of the cell and the level of growth factors that are being produced by neighboring cells. If a cell is proliferating rapidly and the level of growth factors in the environment is low, these two pieces of information are a pretty good indicator that a cell poses a cancer risk. On the other hand, if a cell is proliferating rapidly but this proliferation is in response to growth factors in the cellular milieu, this probably means that the cell is dividing to do something useful for the body, like helping to heal a wound or grow during development. By looking at the proliferation rate and the nearby growth factors, the genetic circuit can make a more accurate assessment of the risk a cell poses to the body, strictly targeting cells that are potentially cancerous.

Having two criteria can enable the genetic networks around cancer suppressor genes to make better decisions, and having even more than two criteria can make those networks more accurate—as long as those pieces of information are being combined in an "intelligent" way that helps distinguish misbehaving cells from normal ones. For example, if a genetic network around a tumor suppressor gene can integrate information about the proliferation rate and growth factors with additional criteria, like the cell's level of DNA damage, the metabolism of the cell, and the presence of survival factors, then the decisions will be more accurate. Using more criteria means that a genetic network can, in principle, better distinguish between a normal cell in unusual conditions and a cancer cell that is behaving inappropriately.

The increased accuracy that comes with integrating multiple sources of information may be why genetic networks around tumor suppressor genes like *TP53* are so complex and connected to the genetic networks that run the cell. This interconnectedness allows the tumor suppressor genes to keep tabs on the whole system and integrate information from all of these different aspects of cell function and physiology. Increasing the accuracy of the cellular decisions requires bringing all the information together in a central node (like *TP53*, for example) where it can get integrated. If this information were processed in separate circuits, then cells would simply have a threshold for "sounding the alarm" if the proliferation rates were too high, and a separate threshold for the alarm if there are damaged proteins or any other potential cue that the cell is a threat. This might be helpful to some extent, but it would miss out on the power of using the combination of the criteria to make a better decision. If, however, multiple criteria can be brought together so the cancer alarms only sound if proliferation rates are high and the cell is producing damaged proteins, then this can make a cancer detection/suppression system more accurate.

This may be why the gene *TP53* is paradoxically both so central and so vulnerable a node in our cancer suppression systems—it has to be central in order to integrate the metaphorical smoke and heat to decide if the combination of the two is problematic. A fire alarm that sounds if either smoke or heat is present is not nearly as good as a fire alarm that sounds when the combination of smoke and heat is present at levels that are likely associated with a fire. Even with access to many criteria for decision-making, it is still a major challenge to create a decision-making system that can distinguish between fire or no fire, guilty or innocent, cancer or no cancer. This accuracy requires a decision system that responds to the right combination of cues indicating danger and doesn't respond to the combination of cues that are associated with a normal or safe state of the system. It's not enough to simply have access to all this information. We have to have a way of processing that information to make the right decision about a potential threat.

The more complex an organism, the more cellular "decision-making" in the form of regulatory systems like *TP53* is required to function properly. When regulatory systems are more complex, it can also be easier to find loopholes and ways of circumventing the rules (the tax code is one example). The more rules there are for how cells should behave and interact, the more complex the signal detection system for cellular cheating will have to be in order to accurately identify cheaters. Thus, multicellular bodies are under selective pressure to leverage cellular computation in genetic networks to find and eliminate potentially cancerous cells—because cancer cells are always evolving to find loopholes in our bodies' cancer suppression systems.

And so we return to the conundrum of *TP53*: Why do we have this central processing system for detecting cellular misbehavior instead of a system with more redundancy? *TP53*'s strength may be its centralized nature. If this process were more distributed, it would be easier for cheating cells to sneak by since the "incriminating" bits of information might be scattered about in several different networks. Having all the information flowing through the *TP53* node may be a way to centralize the processing of all of the information that can help distinguish between a harmless cellular mistake and a sign that a cell is turning cancerous. And by incorporating many criteria into this decision, centralized cancer suppression systems like *TP53* may be able to avoid some of the trade-offs between false alarms and misses.

Looking at the complexity of our cancer suppression systems and the intricacy of the information processing that happens in genetic networks such as those around *TP53* suggests that we need to rethink some fundamental assumptions about cells. Cells are not simplistic input-output machines—they are complex information processing devices that gather and respond to a multitude of signals to "decide" what to do next: whether to divide, repair DNA, self-destruct, or something else entirely. These cells also manage to work together seamlessly to share information within the cellular neighborhood and with the immune system to keep would-be cancer cells at bay.

We are cooperation incarnate, but also intelligence embodied. Our cells are processing and responding to information every millisecond to make us who we are and keep us cancer-free. Our bodies and the cells within us are smarter than we think—and they deploy a network of cellular intelligence (that we are completely unaware of) to detect and suppress cellular cheating. This cellular intelligence is poised to begin monitoring for cellular cheating the moment we are conceived, and it continues throughout our lives. Without constant monitoring and responding to cellular cheating, we wouldn't be able to develop normally, let alone survive long enough to reproduce. Multicellular bodies need cellular cheater detection to escape the fate of becoming evolutionary dead ends. It is an essential part of who we are as multicellular organisms.

Our cells—our whole bodies—are processing huge amounts of information to keep us cancer-free. It's not just our brains that are processing information to help us survive and thrive. It's every cell in our bodies, constantly monitoring itself and policing its neighbors to keep cellular cheating at bay and make us cooperative cellular societies.

The level of cooperation that is achieved by multicellular bodies vastly outshines that reached by human beings. We are a highly cooperative species, capable of working together in large groups and conducting many amazing feats of technology and engineering. Yet our accomplishments pale in comparison to the feats of biological engineering and information technology that cells in our multicellular bodies achieve every minute of our lives, maintaining the complex cellular cooperation that is necessary for us to stay alive. Our cells build (and constantly rebuild) our bodies by proliferating, expressing genes, and making proteins that create both the physical infrastructure of "us"—our cells and the extracellular matrix between them—and the informational infrastructure that makes us function.

We can think of our cells as possessing a form of collective intelligence. Like an ant colony that regulates its temperature or resource foraging through the interactions of the ants, without any one ant understanding the goals of the colony, the cells of our body regulate our temperature and eating behavior without any one cell "knowing" what the organism's goals are. The cells in our body use collective intelligence to accomplish an astounding level of cooperation and conflict suppression every day. This collective cellular intelligence is the basis for both multicellular cooperation and cellular cheater detection, and it is what keeps us alive and thriving from the moment we are conceived and throughout our lives.

4

Cancer from Womb to Tomb

We are born with cancer. We die with cancer. We live our lives with cancer. Cancer is a part of our lives, from womb to tomb. Even if a different disease kills us, we are likely to harbor microscopic tumors while we are on our deathbeds. And many of us live long and happy lives with cancer, carrying tumors along with us as we go about our days. Those of us who are lucky enough not to carry tumors are still hotbeds for cancer mutations and precancerous growths.

Look at the skin on the back of your hand. What do you see? Is it uniformly colored? Is it speckled with freckles? If you are old enough to be reading this, you are likely to have precancerous mutations on the back of your hand. Freckles, moles, skin tags, and even scars all have precancerous mutations (like mutations in the cancer suppressor gene *TP53*). Many cells with precancerous mutations don't even look different from your normal skin. In a study of normal sun-exposed skin (from the eyelids of four individuals), researchers found that these seemingly healthy cells had two to six mutations per million bases, similar to the mutational load found in many cancers. More than a quarter of apparently "normal" cells in this study were carrying potentially cancer-causing mutations. Yet these cells kept acting like normal cells, maintaining the functions of the epidermis (the outer layer of skin). They didn't appear cancer-like aside from having a lot of mutations and being an expanding population of cells. Another study of normal exposed skin (from the forearms of seven individuals) found many *TP53* mutations, and estimated that approximately 0.24 percent of sun-exposed cells were acquiring *TP53* mutations each year. This means that, in the few seconds it takes you to walk from your car to your front door, you accumulate more *TP53* mutations than you could count on both hands.

As we saw in the last chapter, cancer suppression can come with costs like premature aging. This is part of why organisms have not evolved to suppress cancer completely. In fact, our susceptibility to cancer is linked to many traits that allow us to survive, thrive, and reproduce, such as fertility, the body's healing properties, and our ability to fight infection. This enables some germ line mutations that contribute to inherited cancer risk (like *BRCA* mutations) to persist in human populations, despite the costs to health and longevity.

In this chapter I will explore how evolution at the organism level and evolution at the cellular level interact to shape cancer suscep-tibility over our lifespans. Natural selection at the organism level hasn't eliminated cancer susceptibility because of various constraints and trade-offs with other traits. In addition, natural selection among cells inside us happens throughout development, shaping who we are and how susceptible we are to cancer. Even while we are still in the womb, cells in our body are competing with each other and evolving, contributing to susceptibility to childhood cancers. And even though our body's ability to suppress cell competition and evo-lution within us weakens as we near the end of our lives, it is not completely absent.

The Pit of Chaos versus the Swamp of Stagnation

Imagine yourself getting ready to walk a tightrope. Below you, on your left, is a vast pit of chaos—a hot mess that will turn you into an uncontrolled, proliferating mass if you fall into it. On your right is a freezing and stagnant swamp that will paralyze you and swallow you alive, should you fall into it. Whether you realize it or not, you have already successfully traversed this tightrope. Your embryological development, from a single cell to trillions of cells, required it.

If the balance tips too far to the left in this tightrope walk of embryological development—allowing for too much cellular freedom—then the embryo falls into a pit of chaos. Too far in this direction and you have a disorganized mass of proliferating and invading cells. If the balance tips too far to the right—keeping too much control over cell behavior in the developing embryo—then it falls into a swamp of stagnation. If cells lack the ability to proliferate and move around during development, the body would be only a tiny ball with no nervous system and no reproductive system.

Cellular freedom is the root of cancer. If cells have too much freedom, cheating cells can prosper. And if those cheating cells prosper, they will leave more cellular descendants and grow in frequency. This is the same process of natural selection favoring cheating cells that I explored earlier. Our cancer suppression mechanisms help to keep cellular behavior under control and stop somatic evolution in its tracks. Greater cellular freedom means more opportunities for cheater cells to survive and thrive; more cellular control means fewer opportunities for cheater cells—but this comes with a cost.

Greater control over cellular behavior—in the form of cancer suppression mechanisms like the gene *TP53*—can protect us from cancer by slowing down and even shutting down the evolutionary process that favors cheating cells. But too much control can be detrimental to our health and viability. Many important systems that help us survive and thrive require cells to do things that are "cancer-like," including proliferating rapidly, moving around the body, and invading tissues. For example, if you get a cut, the process of wound healing involves cell proliferation and movement to cover the wound. If there was too much control on cell behavior, it would be impossible for the cut to heal. Too many constraints on cell behavior would similarly compromise fertility, make it impossible for our tissues to renew as we age, and even leave us more vulnerable to infection.

The costs of too much cellular control are very clear even during embryological development, when proliferation and cell movement are required in order to make us viable at all. The body has to suppress cancer well enough that cells don't replicate out of control in the womb. But it also has to allow enough freedom for the cells to move around so that the body can develop properly. It's amazing that any of us get out of the womb alive.

With every cell division—starting from the first division of the fertilized egg—comes a chance for mutations to creep into our DNA. And during the earliest phases of life in the womb, cells are constantly dividing, moving around the developing body, and invading existing tissues. This tissue invasion establishes our nervous system, circulatory system, and reproductive system—all the systems we need to be functional beings. How is it that we go through all this rapid cell proliferation and invasion without succumbing to cancer before we are born? And how do most of us stay cancer-free through our reproductive lifespans and often even into old age?

A developing embryo meets all the criteria for evolution via natural selection. It is a population of cells that vary in terms of genetics and epigenetics. These differences among cells are heritable and contribute to differential rates of cell division. Evolution among cells begins as soon as we begin dividing from a single cell and start to form a population of cells that makes up our developing bodies. This population of cells can change over time as some cells die, some survive, and some produce more progeny than others. We develop normally because cancer suppression mechanisms start working in the womb, keeping evolution under control during these critical stages and shaping it in directions that allow for our tissues and organ systems to develop.

Ideally, embryological development is a controlled evolutionary process that results in a cooperative multicellular society in the end. Each of our genomes contains the instructions for building, maintaining, and governing a cellular society of thirty trillion. The cells in the developing body grow, divide, organize, and create an astoundingly cooperative multicellular society. But as this cellular society is growing and maturing, it faces the problem of cheating from within. As I've noted, cellular cheaters can exploit multicellular cooperation and expand in the population of cells in the body, sometimes leading to cancer.

If we could look inside the womb at a newly fertilized zygote, we would see the zygote rapidly dividing as it makes its way toward the uterine lining where it will soon burrow in, attaching itself to the maternal blood supply. This tiny ball of cells is vulnerable as it continues to divide and differentiate to make a viable body. Cancer suppression mechanisms must be strong enough to keep the developing body cancer-free, but also permissive enough to make a viable fetus. Returning to our tightrope analogy, this ball of cells has to maintain a precarious balance with just enough cellular freedom to resist falling into the swamp of stagnation, but not so much cellular freedom that it falls into the pit of chaos. If the balance between cellular restraint and permissiveness is not exactly right, we may fail to grow or grow too quickly, or our tissues might end up in the wrong places, compromising our development and continued survival. Too much cellular evolution and we succumb to cancer before we are even born. Too much control over cell proliferation and migration and we may pay an equally high price: stunted growth and failed development.

At this early stage, disruption to the cell proliferation controls can be disastrous for building a functioning multicellular body. For example, a cell with a mutation in both copies of the cell cycle control gene *TP53* could divide out of control, generate progeny who divide out of control, and eliminate any possibility that the embryo can develop normally. Almost half of conceptions are estimated to fail, 80 percent of which fail before a pregnancy is detectable by standard clinical measures. Many of these failures are known to be associated with chromosomal abnormalities arising during meiosis and fertilization, but some of these failures are likely the result of cancerous mutations that arise during embryonic development and affect development. As we will see, sometimes chromosomes get rearranged as cells are dividing to create all the cell types and tissues of the body. This process can contribute to childhood cancers like certain leukemias.

Development can also fail as a result of too much cancer suppression if cellular behavior is overly constrained, not allowing proper development. During the very early stages of development, cancer suppression mechanisms have to be on high alert to make sure that rapid proliferation and invasion don't get out of control and lead to cancer. Cancer suppression is a much more significant challenge in the womb than it is later in life: Instead of dealing with maintaining an adult body in a relatively steady state of cancer suppression, embryological development requires suppressing cancer-like processes while the body is in a rapid state of growth. At the same time, too much control over proliferation and invasion could interrupt development and lead to an inviable embryo. Given all of these challenges, it is truly astounding that all of us survived these risky months of development in the womb.

Imagine you are walking on the tightrope of development again. In your hands you have a long balancing stick with a bucket hanging on each end. The buckets are filled with something—you're not sure what. But you know that if the bucket on the left is heavier than the one on the right, you will start leaning toward the pit of chaos. If the bucket on the right is heavier, you will lean toward the stagnant swamp. Just the right balance between them must be achieved in order for you to walk across the tightrope without much risk of falling (figure 4.1).

If you could dump the contents of these buckets out and examine them, you would see that they are filled with gene products—proteins made by genes that perform various functions in the body. (Recall, all genes affect our bodies by creating gene products—proteins that perform functions in the cell and in the body as a whole. This process of genes creating proteins is called gene expression.) The bucket on the left is filled with growth factors, survival factors, and other gene products that help cells proliferate and move around the body—these gene products can contribute to cancer, tipping the balance toward chaos. The bucket on the right is filled with gene products that keep cells under control, like the protein p53 (made by the gene *TP53*) that I talked about in the last chapter. Gene products like p53 help suppress cancer by monitoring cells for signs that they have gotten out of control and wrangling them back or prompting them to self-destruct if they are too far gone. The particular balance of these gene products is a key factor in determining whether a body can survive and develop into a normal organism, keeping cancer at bay in the process.

FIGURE 4.1 Successful development requires balancing between too much control over cellular behavior (which can lead to stagnation and failed development) and too much cellular freedom (which can lead to cancer). Development is like walking a tightrope with buckets on either side that are filling with gene products that either tilt you toward more cellular control, risking falling into a swamp of stagnation (on the *right*), or tilt you toward more chaos, risking falling into a pit of cancerous chaos (on the *left*).

The idea of the tightrope walker carrying buckets of gene products helps us see how proper development is a matter of balance between the expression of genes that keep cancer at bay and the expression of genes that promote growth and invasion. It can also help us see that trade-offs underlie many aspects of cancer susceptibility. The ability of our cells to proliferate and move around our bodies is essential for proper development and maintenance of our bodies. But the gene products that make this possible also contribute to our cancer risk—they tip us toward greater susceptibility to cancer (unless of course they are counterbalanced with effective cancer suppression systems). In reality, gene products interact with one another and the cells around them in a complex network with a myriad of positive and negative feedback loops that help keep everything in balance. But for the purposes of understanding how cancer risk is shaped by our development, the tightrope image focuses our attention on the importance of the trade-offs that underlie both cancer susceptibility and cancer control.

Your Mother and Father Are Battling It Out inside You

As your body develops from a single cell to the trillions of cells that make up the body of your adult self, the genes from your biological mother and father are waging quiet battles inside every cell of your body. The genes you inherited from your mother are producing factors that help control growth, filling the bucket on the right and tilting the balance toward cellular control. The genes you inherited from your father are producing factors that spur on growth, filling the bucket on the left and tilting the balance just a little bit more toward chaos.

Why would genes from your mother and father work at cross-purposes? And how is it even possible for genes from your mother and father to behave so differently inside you?

Within each of our cells, there are two copies of each of our twenty-three chromosomes, one from each of our genetic parents. Amazingly, many of these genes seem to "remember" whether they came from your mother or father: they can be maternally imprinted or paternally imprinted. This imprinting happens through epigenetics—changes in the molecules on and around the gene make it more or less likely its DNA will get expressed. This means that genes may be expressed or silenced (in other words, they will make proteins or not) depending on whether they came from your mother or father. And during development the genes on the chromosomes from your mother make proteins that keep growth under control, whereas the genes on the chromosomes from your father make proteins that spur growth—all regulated by the imprinting on your maternally and paternally inherited chromosomes.

Not only does the developing body have to hit the right balance between the production of factors that spur growth versus those that enforce control, it has to do this while the genes from your mother and father are producing factors that could at any moment throw off that precarious balance and plunge the developing body into a pit of cancerous chaos or a stagnant swamp of failed development. As the developing body traverses the tightrope, either paternal genes or maternal genes can throw it off balance.

Genes from your mother and genes from your father can behave differently while you are developing in the womb, but why would they? Shouldn't maternal and paternal genes cooperate to create the healthiest possible offspring? Where does the conflict come from that leads to this battle between the gene products from Mom and those from Dad?

To understand why maternal and paternal interests differ during fetal development, we have to go back to some basic evolutionary theory, the theory of parental investment. Your biological mother and father are genetically distinct entities. This means that their evolutionary interests are not be completely aligned. Sure, they aligned their evolutionary interests to a large extent by having you—their shared offspring—but their interests are not perfectly aligned. This maternal-paternal conflict arises from the fact that humans are not (evolutionarily speaking) entirely monogamous. In a species that was completely monogamous, maternal and paternal interests would be completely aligned because each parent would have no other offspring with anybody else. But humans have diverse mating and marriage patterns, including simultaneously having multiple mates (in the cases of polygyny and polyandry), serially having multiple mates (including serial monogamy, a common pattern for modern Western humans), and sometimes lifelong monogamy as well. So during our evolutionary history, parents often had previous offspring with somebody else, or might have future offspring with somebody else. These diverse mating systems have shaped our biology, from the physiology of pregnancy to our susceptibility to cancer. Let's take a deeper look at how our evolutionary history of being placental, not-entirely monogamous mammals has led to conflict in the womb over gene expression that impacts our susceptibility to cancer.

Milkshakes and Monogamy

To see how this conflict over maternal resources works, let's consider the classic milkshake. In the 1990s, the theoretical evolutionary biologist David Haig proposed what has come to be known as the "milkshake model," which asks us to imagine a mother who has bought a milkshake for her children to share. Here I'll take a version of that model and show how it plays out in the evolutionary logic that underlies conflict in the womb. (I realize this model compares mothers to milkshakes. It goes without saying that mothers are so much more than milkshakes. But mothers can and do directly feed their offspring, and this metaphor is a way to think about how conflict can arise over the allocation of these maternal resources.)

Let's imagine a mother who buys a rather large milkshake for her hungry children (for the purpose of this story, let's assume they do not all share the same father). First, the mother gives the milkshake to her oldest child, who sips a bit. Then she gives it to the next oldest child, then the next, proceeding in that way until every child has had their chance. How much of the milkshake do you suppose is left for the youngest child? Or for the mother by the time the milkshake makes its way back to her? Of course, this will depend on how restrained or ravenous the children are when they take their sips. With the typical voracious ingratitude of many children, there may be little left. But if the children are more restrained, sipping only a moderate amount of the milkshake before passing it on, there is likely to be enough for every child, perhaps even with a sip or two left for Mom.

From Mom's perspective, it's best if the children are more restrained and share equally. The offspring, on the other hand, prefer to have more than their fair share while still leaving their siblings with enough to survive. In Haig's model of this conflict, it is the paternal evolutionary interests that are at work behind the more voracious children (regardless of whether those children are male or female), and it is the maternal evolutionary interests that are behind the restraint in the children who leave enough for their subsequent siblings.

This milkshake dilemma parallels what happens in the womb over the course of several pregnancies. The milkshake represents the store of maternal resources (for example, limiting nutrients for fetal development), and each offspring having his or her turn to suck down part of the milkshake is analogous to the time spent in the womb, where the offspring have access to maternal resources via the placenta. The ravenous versus restrained nature of each child represents the intensity of resource extraction and rapidity of growth of the fetus in the womb.

If a species is completely monogamous, both the mother and father rely on the maternal resources of the mother to feed all the offspring. In the case of complete monogamy, the father's perspective and the mother's perspective are the same: there is just a single milkshake from which all offspring can drink. But if the species is not monogamous, then the father is not relying on resources from just one female for feeding his offspring—in other words, there is more than one milkshake for the father's offspring to drink from. When the father can have multiple reproductive partners, the costs of depleting the milkshake are not as high for the father as they are for the mother.

When a species is nonmonogamous, there are differing evolutionary interests between the mother and father. These conflicting evolutionary interests raise the possibility of conflict over how to invest in offspring, creating a tug-of-war: from an evolutionary perspective, your mother would "prefer" that your father invest more in you so *she* can save more resources for future offspring, whereas your father would "prefer" that your mother invest more in you so *he* can save more resources for future offspring. This doesn't mean that your mother or father is consciously trying to extract more resources from the other. We are simply using a bit of adaptationism here, to see how maternal and paternal fitness interests play out in this complex situation.

The gene expression of a given child affects how hard and fast that child extracts resources from the mother. Because all of the offspring that are carried by one mother must drink from the same metaphorical milkshake, if one offspring consumes a lot, that will leave less for future siblings. It is in the mother's evolutionary best interest to grow a healthy but not too exploitative baby in her womb. But it is in the father's evolutionary best interest to grow a somewhat exploitative baby in the mother's womb because he is not the one who will have to bear the costs of gestating a resource-hungry baby.

Amazingly, this conflict plays out within each cell and within every fetus as it develops from a single cell.

This conflict began with the evolution of internal gestation about one hundred million years ago, and it has continued through the evolution of placental mammals. Mammals such as human beings have a disposable organ designed only for extracting resources from our mothers while we are in the womb: the placenta. It is like a huge, multipronged, invasive straw that burrows into the mother's uterine lining and then sucks up resources to feed the growing fetus. The placenta is genetically part of the fetus. It develops from the same cells of the conceptus (the bunch of cells that come from the zygote, including the embryo and extraembryonic structures) that give rise to the fetus. But the cells that make up the placenta are different from the rest of the cells in the conceptus. Rather than participate in the great choreographed dance of fetal development, they follow a different course. Of all the cells in the conceptus, they are the first cells to differentiate, becoming trophoblasts, which are specialized for invading the lining of the uterus and setting up a resource transfer station to deliver nutrients from the maternal bloodstream to the developing fetus. Not surprisingly, paternal genes express factors that make the placenta larger and more invasive, whereas maternal genes inhibit placental invasion.

The placenta can give the offspring a major leg up when it comes to getting a larger share of maternal resources, leaving less for future siblings. This doesn't necessarily mean that the placentas of later-borns will be smaller than those of earlier-borns. In fact, if the maternal system is more depleted for later-borns, we would predict that the placenta of later-borns will need to actually be larger and more invasive to extract resources. Indeed, a study from the 1950s found that placentas of later-borns tend to be bigger than the placentas of earlier-borns, suggesting that later-borns are building a larger and stronger "straw" to suck up the remaining maternal resources.

Once the resource transfer station of the placenta is set up, maternal and paternal fitness interests do not "agree" about the optimal level of resource transfer. How do these divergent fitness interests play out? Through different gene expression from genes that came from the mother versus those that came from the father. The mechanism that underlies this is the maternal and paternal imprinting of genes that I discussed in the last section. Paternal genes express factors that increase the transfer of resources to the fetus, whereas maternal genes reduce resource transfer to the fetus. Somehow, even with this push and pull of genes working at cross-purposes, bodies can develop into normal little babies.

We can see the dramatic effects of maternal versus paternal gene expression during development by looking at studies of other species. For example, researchers have manipulated the epigenetics of mouse embryos and discovered that if both copies of genes express the "maternal" growth-suppressing gene products, the result will be a tiny mouse; on the other hand, both copies expressing "paternal" growth-promoting gene products leads to a huge placenta. In fact, much of what we know about genetic imprinting and fetal growth comes from studies of mouse placentas. In these studies, researchers have found that paternally expressed genes contribute to the production of more growth factors and greater placental invasiveness, whereas maternally expressed genes do the opposite.

Given what we know about conflict between maternal and paternal genes, we would expect paternal gene expression to dominate in the placenta. Studies of hybrid offspring of horses and donkeys provide a window into gene expression in the placenta. A male donkey and female horse produce a mule, whereas a female donkey and male horse produce a hinny. Researchers looked at the gene expression in the placentas of these hybrid animals to see whether the genes of the father (i.e., the donkey genes in the mule and the horse genes in the hinny) were more likely to be expressed. As predicted, gene expression in the placenta (but not the fetus) was dominated by paternally expressed genes. This suggests that the placenta's invasiveness and promotion of growth are driven by paternally expressed genes.

The growth-promoting nature of paternally expressed genes has implications not just for understanding how gene expression shapes cancer susceptibility in the womb, but also for susceptibility to cancer later in life. Oftentimes placental genes driving growth and invasion that should be silent later in life are re-expressed in cancer.

Tensions between maternal and paternal gene expression may resurface later in life, which can lead to greater cancer susceptibility long after development is complete.

Rapid growth and invasiveness are two of the cell phenotypes that define cancer. Both of them are part of our cellular repertoire and are the result of paternal fitness interests. It's not that paternal fitness interests favor cancer—though paternal evolutionary interests can favor cellular phenotypes that are more cancer-like: more proliferative, more invasive, and better able to extract resources from the host.

In the tightrope walk of embryological development, maternal interests lean a little more toward the right—a little less growth and a little more control. Paternal interests, on the other hand, encourage development toward the left—a little more growth and a little less control. This conflict has pushed both sides to evolve in response to one another, escalating the "effort" they put into achieving their preferred outcome. This situation is often referred to as an arms race, since both sides respond to the other, escalating their investment to win the conflict. This particular arms race leads to tremendous inefficiency and wasted effort in the form of gene products that work against each other. For example, maternally expressed genes produce antibodies that can bind to and inactivate growth factors that are produced by paternally expressed genes. In theory, both sides could scale back their gene expression and achieve the same outcome at a much lower cost. But they don't. This makes no sense from the perspective of what is optimal for the fetus itself. But it makes perfect sense if we consider the conflicting maternal and paternal fitness interests that play out inside every cell in the developing body of that fetus.

Another consequence of this escalated conflict between growth and restraint is that the buckets on both sides have been filled under the "expectation" that the other side will be filled as well. So, when something goes wrong, it can go very, very wrong. For example, if there is a mutation in a gene that is usually maternally expressed, there may not be enough maternal gene products in the bucket on the right to balance out what the paternally expressed genes are putting into the left bucket. This can lead to a negative outcome for both maternal and paternal fitness interests, despite it looking like paternal interests are winning. In fact, a whole suite of syndromes is associated with mutations or deletions of genes that regulate maternal and paternal imprinting of growth and growth suppressor genes. If paternal gene expression dominates (as a result of mutations that disrupt the normal epigenetic regulations), then this can lead to syndromes such as Beckwith-Wiedemann, which is associated with rapid growth in the womb, large size as a child, and an increased risk of cancer.

Despite the fact that we have evolved to be quite functional at the organism level, multicellular bodies like ours are not completely optimized for organism-level cooperation. Conflict still exists within us—among cells and even within them, as is the case with conflict between maternal and paternal gene expression. During fetal development, our bodies expend metabolic energy both on producing growth factors and also on shutting down those growth factors, leading to a dead loss of resources because of this conflict—and a precarious vulnerability if those processes get disrupted and the system gets off-balance. Our growth and development in the womb are not fully optimized for our evolutionary best interests; they are a dynamic compromise between the evolutionary interests of our maternal and paternal sides.

Luckily, the tension between maternal and paternal interests wanes as we mature. Your body plan (where your tissues and organs reside) is basically complete, and the balancing act between cellular control and cellular freedom gets easier as you grow up. Nevertheless, your body faces many trade-offs when it comes to its susceptibility to cancer as you grow to reproductive age, keep your tissues fresh and renewed, heal wounds, and fight infections. Even fertility and being attractive to the opposite sex can influence cancer risk. These abilities are essential if we hope to survive, thrive, and successfully reproduce. They may be associated with an increased susceptibility to cancer, but evolutionarily speaking, the fitness benefits outweigh the costs.

After you are born, you set out on another tightrope, again balancing similar kinds of gene products against each other long enough to successfully reproduce. But this time the risks of too much control over cell behavior have changed. In the womb, too much control over cell behavior can lead to stagnated development. Once outside the womb, too much control over cell behavior carries different risks: if you fail to allow your cells enough freedom you might be more likely to succumb to infection, you might fail to successfully reproduce, or you might age much more quickly. So as you are walking the tightrope of life, your body must effectively balance the trade-offs between cellular control and cellular freedom, keeping cancer at bay while allowing your cells to do everything they need to do to get your genes to the next generation. Sometimes, suppressing cancer too much can negatively affect fitness-enhancing traits.

Growth and development are inherently risky propositions from the perspective of cancer susceptibility. If rapid growth can increase the likelihood of cancer, then shouldn't our bodies grow as slowly as possible? Alas, there are many benefits to being large.

From an evolutionary perspective, one of the greatest benefits of being a large organism is the ability to reproduce successfully. In order to reproduce, individuals must be reproductively mature. Reaching reproductive maturity requires growth, but growing too quickly can be risky. Growing quickly can mean sacrificing essential repair work on DNA along the way, which leaves the organism more vulnerable to cancer.

DNA repair takes time, and so there is a fundamental and unavoidable trade-off between repairing DNA and replicating quickly. Tumor suppressor genes like *TP53* and *BRCA* produce proteins that help control the cell cycle, making the cell stop to fix damage before going on to replicate the DNA and divide. This slowing down of the cell cycle (and ultimately the organism's growth rate) for DNA repair is one of the ways that cancer suppressor genes help to protect growing bodies from cancer. If cells don't properly regulate proliferation, growth, and DNA repair, then the resulting damage to DNA may not get fixed, allowing mutations to propagate. It can take decades for cells with cancerous mutations to proliferate and grow into cancers, so mutations that occur early in development can have reverberating effects over the remaining lifespan.

Once we are fully developed and have reached our adult size, our tissues go into maintenance mode. Rather than growing the body, they simply maintain the body. This means that some of the risks associated with fast growth dissipate. But the risks associated with cell proliferation do not completely go away—maintaining our tissues requires continued cell proliferation because the cells in our body are constantly dying and needing to be replaced, not to mention needing to proliferate to heal wounds. This perpetual turnover can continue to increase our risk of cancer even after we have reached adult size.

The Cellular Fountain of Youth

Throughout our lives, our cells are continuously dividing to replace and renew the trillions of cells in our tissues. Some organs and tissues, like our skin and the lining of the stomach, renew rapidly, whereas others, like heart cells and neurons, don't replicate very much—if at all—once development is complete. But most of the tissues in our body are constantly renewing cells that slough off and replacing cells that apoptose or otherwise die. This capacity for self-renewal keeps us from aging too quickly, and enables us to heal if our tissues are damaged.

Stem cells are largely responsible for this self-renewal. Cell biologists call stem cells "undifferentiated cells," which means they are more or less "general purpose" cells. They have the same genome as every other cell in the body. What makes them different is the expression state they are in—stem cells remain in a "pluripotent" state, meaning that they can become many different kinds of cells. Stem cells can either continue to divide as stem cells or can differentiate into specialized cells: heart cells, liver cells, stomach cells, immune cells, and so on. We have stem cells in all our tissues that help renew and heal our bodies, keeping us healthy and relatively youthful. Stem cells are beneficial—necessary, even—since they allow us to regenerate our tissues and age more slowly. But having stem cells may also leave us more susceptible to cancer because stem cells can replicate more than normal cells can.

One example of how these undifferentiated stem cells can contribute to cancer risk is the increased risk of breast cancer in women
who have a longer period in life before their first pregnancy. During
the first pregnancy, stem cells in the breast differentiate in response
to pregnancy hormones, creating a branching network of ducts and
milk-producing buds (a network that remains in place for future
pregnancies). But before the first pregnancy, these stem cells stay
in an undifferentiated state, awaiting the hormonal cues that will
send them down their paths of differentiation. Women who have
their first pregnancy earlier in life are likely to spend less time with
undifferentiated stem cells in their breasts. The differentiation of
stem cells in the breast is one of the reasons (along with changes
in hormone responsiveness of mammary gland cells) that women
who get pregnant earlier in life have a substantially lower risk of
hormone-positive breast cancer.

Once a former stem cell starts differentiating into a particular
kind of cell, there are only a limited number of divisions possible
before the cell isn't "allowed" to divide anymore. This restriction on
cell division is an important cancer suppression mechanism.

One of the mechanisms underlying this limit to the number of cell
divisions is the shortening of telomeres, sequences of DNA that act as
a protective cap on the ends of chromosomes, with each cell division.
In addition to being a protective cap, the telomeres act as an account-
keeper for cell division. When a cell divides, the telomeres shorten,
and when they are too short a cell cannot divide any longer, a state
that cancer biologists call "replicative senescence." But telomeres can
also lengthen; if a cell produces telomerase, an enzyme that lengthens
the DNA sequences at the end of the chromosomes, the telomeres
lengthen and a cell can divide more than it otherwise would. The
shortening of telomeres is a mechanism that helps protect us from
cancer by limiting the number of times a cell can divide. The production of telomerase is typically highly regulated in normal cells; but,
not surprisingly, cancer cells can evolve to get around this restriction,
extending their replicative lives beyond what is optimal for the body.

Given the role that telomeres play in both renewing tissues and
suppressing cancer, it is not surprising that they are also key players

in the connections between aging and cancer. Research on mice has revealed that the mice that overproduce telomerase exhibit a higher risk for cancer, but if they don't die from cancer, they live longer. Mice that are deficient in producing telomerase or otherwise have shortened telomeres age more quickly, but also have a lower risk of developing cancer. Similarly, when the telomeres of cancer-prone mice are shortened, the risk of cancer goes down. Telomeres can essentially count down cellular divisions and stop a cell from dividing indefinitely, which can provide a big benefit in terms of reducing the risk of cancer. But restricting cell divisions in this way (leaning to the right on our tightrope) can make it difficult to renew tissues.

The tumor suppressor gene *TP53* also has an important role in trade-offs between cancer risk and aging. We saw in the last chapter how *TP53* must "decide" whether a cell poses a cancer risk to the body, and this decision involves trade-offs between two potential errors: a miss (letting a problematic cell survive) or a false alarm (killing a healthy cell). (This is, of course, an oversimplification, but one that helps us see the underlying structure of the problem.) Killing healthy cells takes those cells out of the population and eventually depletes the renewal capacity of tissues.

Experiments in mice with enhanced *TP53* activity can help us see how this works. When *TP53* is constitutively expressed (meaning that it is in the "always on" position, constantly producing the p53 protein), mice have lower cancer risk but age more quickly. Interestingly, when mice were given an extra copy of *TP53* that was not in the "always on" position, but instead was regulated normally (meaning that it only got turned on when it was needed), mice had lower cancer rates yet did not age more quickly. With the tightrope in mind, the protein p53 helps to prevent the organism from tipping too far toward cellular chaos (cancer) but potentially at the risk of tipping toward too much control (premature aging).

This mouse experiment with properly regulated *TP53* is an elegant illustration of how important regulation of p53 expression is to that balancing act. Suppressing cancer is a dynamic process that requires constant updating and information processing by the genetic networks in the cells of our bodies—and this experiment suggests that we may be able to escape from at least some of the trade-offs between cancer and aging through proper regulation of those cancer suppression mechanisms. Suppressing cancer while avoiding some of the negative effects that can accompany it, such as premature aging, requires smart regulation and "decision-making" by genetic networks that keep tabs on cellular behavior.

Time Heals All Wounds, but Hopefully Not Too Quickly

If you get a cut on the surface of your skin, the cells around the cut have to be able to proliferate to create new cells that can cover the wound and rebuild the tissue. These cells also have to be able to move, creating fronts of motile cells that can come together and close the wound. These are the very same abilities that cancer cells use to grow and colonize our body. Being able to heal wounds quickly provides huge advantages for the organism. Not only does it allow us to get back to normal functionality faster, it also decreases the likelihood that the wound will become infected.

Evolution has therefore equipped us with the ability to heal quickly, but this capability comes at a cost: it means that our cells are poised and ready to proliferate and move when they get the signals from the body that it is time to close a wound. And when cancer cells begin "falsely" producing these wound healing signals (like factors that increase inflammation), they are able to get around the normal checks and balances that keep multicellular cells behaving normally. In fact, cancer is sometimes referred to as "the wound that does not heal." Our ability to heal wounds can sometimes get hijacked by cancer cells. Some cancers exploit the signaling systems underlying healing, keeping our tissues in a state of constant inflammation.

Our bodies have evolved to dynamically adjust the balance between cellular freedom and control depending on what we need to accomplish. If we have a wound that needs to heal, the balance tips a little bit more toward cellular freedom. In the tightrope walk of life, the gene products that get produced during wound healing fill up the bucket on the left. But this lean to the left is only temporary, until the wound is healed. Sometimes, cancer cells evolve inside the body to produce factors that tip the balance to the left, making the body more tolerant of cellular misbehavior. And rather than this being a temporary state, as it is during normal wound healing, the cancer cells continue producing these factors that keep the balance leaning toward proliferation. Cancer cells essentially evolve to produce factors that mimic the profile of a wound healing environment and do so indefinitely. This environment gives cancer cells a survival advantage.

Fighting Infection with Somatic Evolution

Our skin is an essential part of our immune system. If it is breached, we become much more vulnerable to infection from bacteria, viruses, and other entities that may try to hijack our bodies for their own fitness interests. When a wound is healing, our innate immune system plays an important role. The innate immune system is the first responder to potential infectious threats. Inflammation is the main tool that it uses to protect us, and cancer can hijack this inflammatory response.

But we also have a more sophisticated immune system that our bodies use to remember previous infections so we can respond more rapidly in the future. This is the adaptive immune system. The adaptive immune system is perhaps our best tool in the arms race against pathogens that might try to evade the immune response. It works by generating slightly different genetic variants of immune cells that can identify novel pathogens. When one of those immune cells finds a pathogen, the adaptive immune system creates more copies of the type of cell by allowing it to proliferate. The end result is that the population of immune cells dynamically responds to the particular threats that the organism is facing. The brilliance of the adaptive immune system is that it essentially deploys cellular evolution within the body to fight pathogens that are themselves evolving. Without it, we would be far behind in the evolutionary arms race with pathogens. The adaptive immune system allows us to mount a response to rapidly evolving pathogens by maintaining our capacity for cellular evolution among immune cells.

The adaptive immune system is like a little bastion of cellular freedom within an otherwise tightly regulated body. It tips the balance toward cell proliferation and the rapid expansion of cellular populations that are specialized for dealing with threats from pathogens. The immune system must strike a balance between too much cellular control, which can increase our risk of dying from infectious disease because the body can't respond as well to incoming threats, and too much cellular freedom, which can increase the risk of immune cancers like leukemias.

Leukemias affect a surprisingly large proportion of children before the age of 15. (Leukemias are among the most common childhood cancers, but the majority of cases of leukemia are diagnosed in adults over the age of 65.) Childhood leukemias are now highly treatable (which may be due to the fact that they are typically genetically homogeneous and therefore don't evolve resistance as readily as more genetically heterogenous cancers—a topic I will return to later in the book). Acute lymphoblastic leukemia (ALL)—the most common childhood leukemia—can occur very early in development when undifferentiated immune cells, called immature progenitor cells, proliferate too much. Although it often originates during development in the womb, it can be triggered by certain patterns of exposure to infectious disease after birth.

As we have seen, leukemias are often caused by translocations— when portions of two genes get put together that should not be next to each other. In leukemias, translocations typically involve two chromosomes swapping chunks of genes. We know that these leukemias originate in utero because researchers studying children with leukemia have gone back to look at blood samples that were collected from the children at birth as part of the routine heel prick procedure (used to screen for genetic diseases such as phenylketonuria). The abnormal translocations were already present in the blood of the newborns later diagnosed with leukemia. Interestingly, approximately 1 percent of newborns have preleukemic clones with these translocations, but only a tiny fraction of them go on to develop clinical ALL. The fact that many newborns with preleukemic clones do not develop ALL shows that susceptibility to ALL does not simply arise from this genetic translocation, but rather that there must be other factors involved.

One of these other factors is delayed exposure to infection in early childhood. When children are not exposed to infection early in life, but are then exposed to a highly infectious agent, this is associated with an increased risk of ALL. Mel Greaves, an evolutionary cancer biologist who specializes in childhood leukemia, has investigated several cases of ALL "clusters"—Greaves traced them back to likely infectious exposures that occurred not long before diagnosis. For example, Greaves investigated a disturbing pattern of leukemia in children in Milan, Italy. Seven children, aged two to eleven, were diagnosed within a four-week period. He discovered that this pattern of leukemia occurred after an outbreak of swine flu and that every child diagnosed with ALL had been infected with the swine flu. He also found that the children who were most likely to get leukemia were those who didn't go to day care at an early age or have older siblings—meaning that they had fewer exposures to infections in general during early development (compared to those children who were exposed to other children from an early age) and likely had less developed immune systems as a result. This lack of early exposure to infections may have left them more susceptible to leukemia when they eventually encountered an infectious agent—the swine flu.

It is likely that susceptibility to leukemia exists because of the evolutionary advantages that come with having an adaptive immune system. The capacity for somatic evolution in the immune system is so beneficial—protecting us against infection—that it outweighs the evolutionary costs of having some susceptibility to cancers of the immune system like ALL.

Fertile Conditions for Cancer

We have seen how our susceptibility to cancer is linked to growth, tissue maintenance, wound healing, and protection from infection. Cancer suppression is also linked to the holy grail of evolutionary viability: fertility and reproduction. Controlling cell proliferation and DNA repair is a good thing when it comes to suppressing cancer. But—sometimes it can negatively affect fertility. One example of a trade-off between fertility and cancer suppression comes from a study of women with mutations in *BRCA* genes, genes that are involved in DNA repair.

BRCA1 and *BRCA2* are two different (and unrelated) tumor suppressor genes that both produce proteins that are responsible for DNA repair, and also play a role in the formation of oocytes (cells in the ovary) and embryonic development. Sometimes, individuals are born with germ line mutations in *BRCA* genes, which can leave them more susceptible to breast and ovarian cancer during their lifetimes because mutations in these genes can lead to faulty DNA repair (*BRCA* mutations are associated with many other cancers as well). *BRCA1* and *BRCA2* are long genetic sequences, on chromosomes 17 and 13, respectively. Because both *BRCA* genes are quite long, there are many potential mutations that can happen, and these different mutations have different consequences for the production of the DNA repair proteins that these genes typically produce. Some mutations in *BRCA* genes can completely disrupt the production of the protein and subsequent DNA repair, other mutations only partially disrupt the production of the protein, and some mutations have no effect on the production of the protein. This means that some *BRCA* mutations are not associated with elevated risk of cancer. This diversity of cancer risk associated with different *BRCA* mutations (often in different ethnic groups or subpopulations) makes clinical management challenging. Not all *BRCA* mutations create health problems—and so more extreme preventative measures like double mastectomy may not be appropriate for all *BRCA* mutation carriers. Sometimes, bilateral mastectomy is performed for women with nonpathogenic *BRCA* mutations, and many of these women never receive genetic counseling that might help them interpret their results and better understand their risk.

Germ line mutations in *BRCA* genes can be passed down from generation to generation. The *BRCA* genes—like most genes in our genomes—are made of thousands of base pairs. This means that there are many possible mutations in the *BRCA1* and *BRCA2* genes, and some of these mutations contribute to our risk of cancer. *BRCA* mutations confer a 65–80 percent risk of breast cancer in women who harbor these mutations, compared to a 12–13 percent risk in the general female population. Women with *BRCA* mutations are often diagnosed with cancer during their reproductive years (about 25 percent of *BRCA1* mutation carriers are diagnosed with breast cancer before the age of forty, and 72 percent are diagnosed by the age of eighty). *BRCA* mutations are not limited to women; men with *BRCA* mutations have increased risk of breast and prostate cancer. Why hasn't natural selection eliminated these harmful *BRCA* mutations from human populations? One possibility is that the risk of breast cancer that comes with some *BRCA* mutations may be accompanied by traits that provide a fitness benefit—for example, a boost in fertility that allows a woman to have more offspring than she otherwise would. Several studies have investigated this connection using information about *BRCA* mutation status and fertility from large databases such as those from the Utah Population Database, which contains health records of millions of women in Utah, going back several generations.

In Utah, all breast cancer diagnoses are recorded by physicians in the state cancer registry, where they can be cross-referenced with other records including family history data (this database can only be used by researchers and the privacy of the records is strictly protected). The family history data often include birth records for the individual's mother, grandmothers, and even great-grandmothers. These data make it possible for researchers to examine whether susceptibility to breast cancer is associated with fertility patterns in a woman's ancestors. And this database with family and medical histories goes back far enough that it includes fertility data from before hormonal birth control became available. This information about fertility rates before birth control is quite valuable to researchers who are looking at potential trade-offs when it comes to cancer susceptibility genes and fertility (because in populations using contraception, fertility overall is lower and so it can be difficult to see an effect).

This amazing resource allows researchers to look for associations with cancer susceptibility genes as they relate to the number of offspring a woman has had. In one fascinating study, researchers found that women with a mutation in *BRCA* genes were more likely to be diagnosed with cancer and had higher mortality than women without it. But what made this study really interesting is that the researchers went back in the Utah Population Database to examine the fertility of female relatives of women who had these *BRCA* mutations—female relatives who lived before hormonal contraception was available. They found that women with *BRCA* mutations had female ancestors that had more offspring—with an average of 1.9 more offspring compared to the ancestors of women without these *BRCA* mutations (for women born before 1930 the controls had on average 4.19 offspring, whereas carriers had an average of 6.22 offspring). This suggests that *BRCA* mutations create a link between fertility and cancer susceptibility, at least in this particular population.

Another study—this time using information collected in a database from central France with more than one hundred thousand individuals—found similar effects of *BRCA* mutations on fertility. Women with *BRCA* mutations in this sample had more children (1.8 more on average compared to controls), were less likely to have no children, and had lower rates of miscarriage. Interestingly, men with the *BRCA* mutation also had more offspring than men without it.

But this link between fertility and *BRCA* mutations does not seem to hold across all populations. For example, a study of women from the United States and Canada did not find a significant relationship between fertility and *BRCA* mutations. This sample was comprised of relatively young people and included women using contraception, potentially making it more difficult to see an effect if it were there. Another study found no evidence of increased fertility in women with *BRCA* mutations, though the researchers did find that women with *BRCA* mutations had more female offspring (almost 60 percent females) than women without these mutations (who had just over 50 percent female offspring). These contradictory results could be due to the fact that the link between fertility and *BRCA* is not generalizable across all populations with *BRCA* mutations. In addition, there are many different *BRCA* mutations, and it may be that only some of these mutations in some populations have a trade-off with fertility. In addition, it could be that some *BRCA* mutations provide a benefit for male fertility—and neither of the studies that failed to find an effect looked at the relationship between *BRCA* and fertility in males. This is clearly an area of active research with many open questions to be answered.

Different Human Populations, Different Cancer Risk Genes

In many populations, the risk of breast and ovarian cancer is linked to *BRCA* mutations, but the particular mutations differ across populations. Some ethnic groups—such as Ashkenazi Jews—have common mutations in *BRCA* genes that point to a common ancestor. These kinds of population-specific risk genes often occur as a result of the founder effect, which can happen when an event—whether it is migration, infectious disease, or even human-caused decimation of a population—leads to a small "founder population" that expands into a much larger population over the course of several generations. Because many individuals in this population share a common (and recent) ancestor, they are more likely to share common genes, including genes that may contribute to a susceptibility to cancer (and possibly fertility as well). Specific *BRCA* mutations resulting from a founder effect have been discovered in many human populations; in addition to Ashkenazi Jews, populations are found in Norway, Sweden, Italy, and Japan. Each of these populations has specific *BRCA* mutations that get passed down from generation to generation through the germ line (i.e., the sperm and/or egg cells). Many of these mutations are associated with increased risk of breast and ovarian cancer—though the risk of cancer varies for the particular mutation and for the particular population.

There are many reasons why different mutations are associated with cancer in different populations. As we've seen, cancer is not caused by mutations alone, but also by an imbalance in gene products that regulate cell behavior. That balance of gene products is influenced by inputs from the environment that regulate gene expression and also by the background of other genes and how they are being expressed. In other words, different populations have different gene products filling the metaphorical buckets balancing growth and constraint. For a person from one population—say, an English woman—a mutation in *BRCA* might be enough to tip the balance toward much greater cancer susceptibility. But for someone from another population—say a Norwegian woman—a mutation in *BRCA* might not have as much influence on cancer susceptibility because it doesn't tip the balance. Not all mutations in "cancer risk genes" like *BRCA* are the same—some confer little, if any, risk of cancer.

Cancer is an ancient disease, but some aspects of cancer are more evolutionarily recent than others. Evolution cannot completely eliminate our susceptibility to cancer, and cancer originated at the very beginning of multicellularity itself. These ideas offer a new way of thinking about inherited cancer susceptibility—rather than thinking only about gene variants like *BRCA* mutations as part of our inherited cancer susceptibility, we can take a broader view as we consider our genomes and long evolutionary history as multicellular life forms as well. Because of trade-offs with fitness-relevant traits and constraints, our susceptibility to cancer has been passed down from one generation to the next, contributing to inherited susceptibility since the dawn of multicellular life on our planet.

Some aspects of this inherited cancer susceptibility are simply part of the legacy of multicellularity. Other aspects of inherited cancer susceptibility are more evolutionarily recent. For example, it's likely that genes involved in placentation contribute to cancer given their role in tissue invasion—and so this inherited susceptibility originated when placental mammals came on the scene. We all carry with us these ancient inherited cancer risks of being multicellular organisms and being placental mammals.

In contrast, the kinds of genes we typically think of as inherited cancer risk genes, like *BRCA* mutations, are not shared by all of us. In fact, it is the variation among humans that lets us see how they contribute to our risk of cancer. Many of these inherited risk genes that vary among humans have arisen more recently, typically in bottlenecked populations where they may have in some cases provided a fertility advantage. But there are other inherited risk genes that are clearly deleterious and evolutionarily harmful, and they exist simply because they have not been around long enough to be selected out of the population.

Invasion of the Trophoblasts

BRCA is not the only gene family that links cancer and fertility. *KISS1* is a gene that produces a protein (kisspeptin) that helps control placental invasiveness, and plays a role in pubertal development. One of the roles of kisspeptin is to suppress invasion of trophoblasts into the uterine lining and inhibit angiogenesis (the creation of a blood supply that provides resources for the fetus). But *KISS1* is also involved in cancer suppression: it helps suppress metastasis in breast cancer and melanoma. This may come as no surprise, given that metastasis and placental invasion have many similarities in terms of the underlying mechanisms. If reproductive tissues are more easily invaded by fetal trophoblasts (the cells that form the placenta), this can make it more likely that a pregnancy will take hold. In order for an embryo to implant and begin extracting resources, it needs to invade the uterine lining. Higher tolerance or even receptivity to invading cells could make a woman more fertile, but can also leave her more vulnerable to invading metastatic cancer cells.

Similar trade-offs between cancer susceptibility and reproduction may exist for males as well. For example, the risk of prostate cancer is associated with higher exposure to testosterone. Testosterone is also associated with greater investment in mating behavior. Higher testosterone levels may facilitate engaging in more short-term mating, but, as some researchers have pointed out, high levels of testosterone are also correlated with higher risk of prostate cancer in the long term.

All of this suggests that, from an evolutionary perspective, the optimal level of cancer defenses may not be as high as you might think. If too much protection against cancer can have a negative effect on survival and reproduction, evolution could favor lower levels of cancer defenses. My colleagues and I were interested in this issue and the trade-offs between reproduction and cancer risk, so we created a computer model that allowed us to look at how the optimal level of cancer defenses differed in various reproductive environments. We wanted to see whether the balance would tip away from cancer suppression if organisms evolved in an environment with lots of reproductive competition, where only the most competitive individuals succeed in reproducing.

We found that the optimal level of cancer defenses was quite low when being more reproductively competitive was crucial for reproductive success. For example, when competition for reproduction was more of a "winner-take-all" situation, with the most competitive individual getting all the mating opportunities, the model predicted that cancer defenses would evolve to be extremely low. Cancer defense only paid off when extrinsic mortality (the likelihood of dying from random causes) was low and competitiveness made little difference to reproductive success (when it was *not* a "winner-take-all" mating system).

Many traits evolve because they directly enhance reproductive opportunities or because they are simply preferred by the opposite sex. These are called sexually selected traits, and include larger body size, and ornaments, such as antlers. Sexually selected traits sometimes require high levels of cell proliferation for the creation of secondary sexual characteristics and ornaments. This need for rapid cell proliferation might tip the balance toward greater cancer susceptibility (like the antleromas, or antler cancers, which I will look at in the next chapter). Some of the most breathtaking biological forms we see in nature—beautiful, colorful markings and enormous antlers—can carry with them an increased risk of cancer. Traits like faster cell proliferation, sloppier DNA repair, and more permissive conditions for conception and/or implantation of an embryo can provide an organism-level advantage in terms of reproductive competitiveness, but may come at a cost in terms of cancer susceptibility.

This link between reproductive competitiveness and cancer could play out through mechanisms like the *KISS1* gene. Earlier, we saw how *KISS1* plays a dual role in suppressing placental invasion and inhibiting metastasis. Interestingly, *KISS1* is involved in many other processes linked to fertility as well, including the production of luteinizing hormone and follicle-stimulating hormone, both essential hormones in the female reproductive cycle.

These findings suggest that being a large, fertile, attractive organism that is investing a lot in being reproductively competitive can sometimes come with hidden costs, like a greater susceptibility to cancer. Of course, this doesn't mean that being more susceptible to cancer makes individuals sexier and more fertile—only that these traits may have trade-offs that make an organism more vulnerable to cancer.

We All Live with Precancerous Growths

Implantation of the placenta in the uterine wall is one example of a crucial function that requires tolerance of invasive, cancer-like behavior. As we have seen, other cancer-like phenomena play a similarly essential role in many important fitness-enhancing traits and cellular activities, like wound healing, which requires rapid cell proliferation, cell migration, and recruitment of blood vessels to feed and rebuild a healing tissue—all characteristics that we see in cancer cells. Some of our susceptibility to cancer is a trade-off for having traits that allow us to develop normally, survive, and reproduce. Cancer is the evolutionary price we have to pay for benefits like being bigger, healthier, and more fertile.

These trade-offs between cancer suppression and other fitness-enhancing traits play out over the course of our lifetimes. They mean that our bodies tolerate some level of cancer-like behavior from our cells so that we can do all the things we need to do to be reproductively viable. As we age, we accumulate precancerous growths. As we have seen, cells don't turn cancerous the moment they get a few mutations. They can continue to behave normally, behaving as healthy, functional cells contributing to our multicellular body.

Even if we don't die of cancer, we almost certainly die *with* cancer—or at least cancer-like growths. Most men will die with (but not because of) slow-growing tumors in their prostates. Many women will die with breast tumors that might or might not qualify as cancer. Most people die with microscopic thyroid tumors. And our skin is constantly acquiring precancerous mutations as a result of sun exposure, wound healing, and other routine exposures. We live with precancerous growth for decades, usually without any problems.

Our bodies develop cancer-like growths, but as long as they stay local, our bodies can keep these growths under tight controls, thanks to our evolved cancer suppression systems. But if they get out of control, invading neighboring tissue and metastasizing throughout the body, they can be deadly.

Cancer is the process of somatic evolution within the body, but our bodies can tolerate quite a bit of somatic evolution and mutational load, keeping it under control while still maintaining the functions of the body. We can live with cells constantly evolving without succumbing to cancer.

Our susceptibility to cancer comes along with normal physiological processes that occur throughout our lives, many of which promote health, enhance reproduction, or protect us from threats. But what happens to our susceptibility to cancer as we pass reproductive age and start to enter old age?

As we age, the power that natural selection has had to shape our bodies starts to diminish. This is because what we do after reproduction typically matters less for our evolutionary success than what we do beforehand. Sometimes people use this principle to argue that there is no selection for cancer suppression after reproduction. But this is simply not true. Selection for cancer suppression systems weakens, but it does not disappear entirely in old age. Humans are unique in that we have extremely long periods of parental investment (our hunter-gatherer ancestors likely invested in their offspring and even grand-offspring for decades after birth). Long-term parental investment means that it is still possible to enhance the reproductive success of your offspring in old age. So there is still some selection for cancer suppression even late in life. I have looked at this very issue with my colleague Joel Brown. Employing a mathematical model, we found that in organisms (like humans) who invest a lot in offspring after they are born, selection for cancer suppression can remain high enough after reproduction to favor cancer suppression mechanisms in old age.

Life expectancies are greater today than they were in ancient history, but many of our ancestors lived to old age as well. Data from modern hunter-gatherers show that humans in conditions similar to those of our ancestors often live past seventy years old. This long postreproductive life means that selection for cancer suppression mechanisms did not completely drop off for our ancestors after reproduction. Nevertheless, we live longer lives today than our hunter-gatherer ancestors did, and we are more likely to experience cancer largely because we are less likely to die of other causes early in life—like accidents and infections.

Modern Exposures Contribute to Cancer Risk

We evolved to suppress cancer in a world that was quite unlike the world we live in today. There were no vending machines, no escalators, no shift work, no cigarettes. The world that our cancer suppression systems evolved for is that of our hunter-gatherer ancestors. Daily life for hunter-gatherers involves miles of walking to collect fruits and berries or hunt for game, climbing up cliffs and trees in search of beehives to collect honey, and often strenuous digging for tubers. Every calorie a hunter-gatherer consumes is hard-won. This is a far cry from modern life where we easily consume many more calories than we need, and we walk much, much less than hunter-gatherers do—even if our step-tracking wearables log an impressive number of footfalls.

In addition to increased risk from more calories and more sedentary behavior thanks to modern conveniences, our life is associated with other exposures like chemical carcinogens (the most important of which being the carcinogens found in cigarettes), higher levels of reproductive hormones (because of better nutrition and, for women, more frequent ovulation), and greater disruption to our sleep (due to artificial light, shift work, and late-night screen use). Over the course of our lifetimes we are exposed to many different substances and events that our hunter-gatherer ancestors never experienced—and these changes in exposures happened too quickly for human populations to have yet evolved better cancer suppression mechanisms.

Cancer itself is an ancient disease, but our modern lifestyles increase our mutation rate through carcinogenic exposures, or by shifting the delicate balance in the body away from cellular control and toward cellular freedom. For example, higher reproductive hormones can shift cells of the body toward faster proliferation, potentially sacrificing DNA repair or other aspects of somatic maintenance that might help protect against cancer. We also live longer than our ancestors did, due to better nutrition and better medical care, meaning that we have more years at the end of our lives during which cancer can emerge.

Our susceptibility to cancer begins when we are first conceived, but there are many contributors to our susceptibility that were set in place long before our parents met, before modern humans evolved, and even before placental reproduction evolved. Our susceptibility to cancer is rooted in our evolutionary past, but it also involves an evolutionary struggle that happens over the course of our lifetimes among the cells inside us. In order to function normally as a multicellular organism, our cells need to be able to proliferate, move, and use resources around them. But these cellular capacities contribute to our susceptibility to cancer. Letting up on cancer suppression can have substantial evolutionary benefits like making an organism bigger or more fertile. This leads to the counterintuitive conclusion that the optimal level of cancer risk for an organism is not zero. If we were to completely suppress cancer, the evolutionary price might be impossibly high.

5

Cancer across the Tree of Life

When Joshua Schiffman's Bernese mountain dog was diagnosed with cancer, Schiffman was incredulous. Schiffman is a cancer researcher and pediatric oncologist—and a cancer survivor himself. The last thing he expected was for his beloved family pet to be a victim of the disease that he studied, the disease that he had grappled with as an adolescent. Schiffman realized that cancer was not just a disease that affects humans. It affects many other organisms across the tree of life.

This experience with cancer in his pet motivated Schiffman to learn more about cancer susceptibility in dogs—and he was surprised to find many similarities with cancer susceptibility in humans. Dogs, like humans, can have mutations in *BRCA1/2* genes that confer greater risk of breast and ovarian cancer. Dog cancers can also have mutations in *TP53*. Mutations in *TP53* in humans results in a syndrome known as Li-Fraumeni, an inherited disorder that makes people much more likely to develop cancer over their lifetimes. Dogs with chronic myeloid leukemias have even been found to have the *BCR/ABL* translocation, the same translocation that I discussed in chapter 3, which is typical of chronic myeloid leukemias in humans.

The similarities between dog cancers and human cancers go beyond genetic risk factors like changes in *BRCA*, *TP53*, and *BCR/ABL* genes. In both dogs and humans, the risk of cancer is associated with larger size. In the last chapter, we saw how growing to a large size, and doing so very quickly, can lead to a higher risk of cancer because of trade-offs between cell proliferation and control over cellular behavior. As organisms walk the tightrope of development, their cells must balance between too much proliferation and too much control in order to develop normally and be evolutionarily fit as adults. And so, all else being equal, we might expect larger organisms to be more susceptible to cancer.

But all else is not equal—at least, when it comes to elephants. Elephants have about one hundred times more cells than we do. Yet their rates of cancer are much lower than ours. In fact, if we look across species we see that larger size does not confer a higher risk of cancer—this pattern of higher cancer risk being associated with large size appears only to hold within species. In the last chapter we saw how being larger can come with increased cancer risk because more cell divisions are required to get to this large size and maintain it. So why, then, is large body size not correlated with cancer risk across species?

In this chapter, I will outline the ways in which an evolutionary approach can help resolve this conundrum, known as Peto's Paradox. I will also take a closer look at some of the reasons why susceptibility to cancer varies across the tree of life, from the simplest multicellular organisms to large and complex forms of life, like elephants. Understanding how cancer affects other forms of life— and how life has evolved to suppress cancer—can give us insights into why we are susceptible to cancer as humans, and guide us to new strategies for treatment and prevention.

Susceptibility to cancer differs across forms of life due to some of the same trade-offs we learned about in the previous chapters: the cost of being big, growing fast, healing wounds, and being reproductively viable. But I will also add another layer to our understanding: evolutionary life history theory, a framework that can help us understand why some organisms invest a lot in cancer suppression mechanisms, whereas other organisms don't. This theory can help explain why some species are particularly resistant to cancer.

I will also look at several curious cases of contagious cancer—from a sexually transmitted cancer in dogs, to a cancer transmitted through face-biting in Tasmanian devils, to a few rare cases of human-to-human cancer transmission. We will see how trade-offs with other fitness-relevant traits—like wound healing and reproduction—shape susceptibility to transmissible cancers just as they shape susceptibility to cancer more generally. We will also see how transmissible cancers have been around since the origins of multicellular life, and how they may have played in important role in shaping many fundamental aspects of our biology, from our immune systems to the evolution of sex.

Cancer across Life

At the beginning of this book, I talked about crested cacti and the amazing variety of crowns, knobby growths, and brain-like formations that can result from disruption to their normal cell proliferation controls. These crested cacti are so intriguing because they are essentially the cactus version of cancer—cells that have escaped from the normal restraints on multicellular behavior and are growing out of control.

Crested cacti are one example of fasciated plants. Plants like crested cacti are literally multifaceted, with growth patterns that often have lots of different faces or facets—hence the term "fasciated." Fasciations can arise when the cells on the growth tip of the plant (called meristem cells) expand from a single tip into a line of cells. As these cells divide they create an expanding band of proliferating cells that grows into a fan-like shape and sometimes even starts folding in on itself to create brain-like growth patterns. This phenomenon of fasciation is not limited to cacti—it happens in many other plants (figure 5.1). Some flowers can become fasciated, leading to the growth of bizarre, elongated blooms. Tobacco plants are

FIGURE 5.1 Plants are susceptible to cancer-like phenomena called fasciations, in which mutations in the growing tips can lead to striking and often beautiful growth patterns. Images from left to right are fasciated as the result of damage on a cypress tree, *Chamaecyparis obtusa* (Anton Baudoin, Virginia Polytechnic Institute and State University; Bugwood.org is licensed under CC BY 3.0); a crested *Casuarina glauca,* which lacks the typical branching structure and instead features a large fan of tissue with dysregulated differentiation (Tyler ser Noche, *File:Starr-180421-0291-Casuarina glauca-with fasciated branch-Honolua Lipoa Point-Maui (41651326770).jpg* is licensed under CC BY 3.0); a mule's ear flower, *Wyenthia helianthoides,* with a normal flower shown on the left and fasciated form on the right (Perduejn, *Mules Ear Fasciated* is licensed under CC BY 3.0); and a "double flower" *Anemone coronaria* (Thomas Bresson, *2014-03-09 14-30-31 fleur-18f* is licensed under CC BY 3.0).

often fasciated, altering their leaf and flowering patterns. Even huge trees such as pines can become fasciated, leading to precariously wide and heavy trunks that expand out into fanlike shapes as they climb to the sky.

Over the years, my interest in crested cacti grew into a fascination with fasciation, which then expanded further into the tree of life, eventually covering all branches of multicellularity. Technically, plants are green algae—they are part of the taxonomic group of green algae that includes everything from giant pines to pond scum.

Cancer and cancer-like phenomena occur in green algae, and—as my colleagues and I have found—every other branch of the tree of life.

There aren't many opportunities in academic life to spend a whole year talking with and working with a group of people who share your general interests—let alone a group of people who are interested in some of the very same questions that you are interested in. I had the privilege and honor to be a part of a working group at the Institute for Advanced Study in Berlin (called the Wissenschaftskolleg, or Wiko for short) where we had exactly this opportunity. Michael Hochberg, a theoretical evolutionary biologist and ecologist studying cancer, organized and convened a working group at Wiko on the topic of cancer evolution. We spent the greater part of our year working on a review of cancer across the tree of life, scouring the literature for reports of cancer and cancer-like phenomena across all branches of multicellular life. We found evidence in clams, insects, all manner of animals, coral, fungi, and, of course, plants (figure 5.2).

We also found that cancer—across all these different species—always involved cheating in the foundations of multicellular cooperation: disruptions to proliferation control, inappropriate cell survival, messed-up cellular division of labor (in other words, dysregulated cell differentiation), resource monopolization, and destruction of the extracellular environment. As I discussed earlier, cellular cheating provides a common framework for talking about cancer and cancer-like phenomena across species. Unlike many other definitions we could use, defining cancer as cellular cheating can allow us to talk about cancer across organisms that have very different underlying biology.

Cancer is typically defined in an animal-centric way, using the criteria of invasion and metastasis. Invasion requires that cells break through a basement membrane, but not all organisms have encapsulated tissues with basement membranes to break through. Not all organisms have circulatory systems that could be used for metastasis either. A more general way to define cancer is by focusing on cellular cheating. This allows us to use a widely applicable set of characteristics across life—those associated with a breakdown of multicellular cooperation.

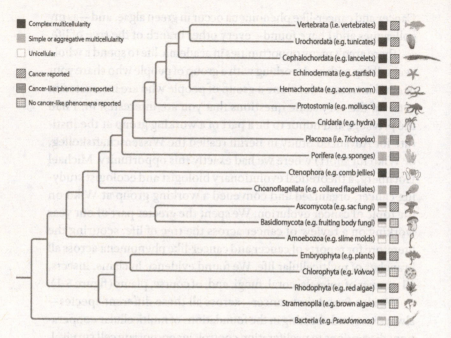

- ■ Complex multicellularity
- ▨ Simple or aggregative multicellularity
- ☐ Unicellular

- ▨ Cancer reported
- ▤ Cancer-like phenomena reported
- ▦ No cancer-like phenomena reported

Vertebrata (i.e. vertebrates)
Urochordata (e.g. tunicates)
Cephalochordata (e.g. lancelets)
Echinodermata (e.g. starfish)
Hemachordata (e.g. acorn worm)
Protostomia (e.g. molluscs)
Cnidaria (e.g. hydra)
Placozoa (i.e. *Trichoplax*)
Porifera (e.g. sponges)
Ctenophora (e.g. comb jellies)
Choanoflagellata (e.g. collared flagellates)
Ascomycota (e.g. sac fungi)
Basidiomycota (e.g. fruiting body fungi)
Amoebozoa (e.g. slime molds)
Embryophyta (e.g. plants)
Chlorophyta (e.g. *Volvox*)
Rhodophyta (e.g. red algae)
Stramenopila (e.g. brown algae)
Bacteria (e.g. *Pseudomonas*)

FIGURE 5.2 All branches of multicellular life are susceptible to cancer. In our review of cancer across the tree of life, we found reports of cancer and cancer-like phenomena (dysregulated differentiation and overproliferation) across every branch of multicellularity. Figure reprinted with permission (Aktipis 2015, licensed by CC BY 4.0).

Many biologists have assumed that invasive cancer cannot occur in plants because plants have cell walls and more rigid tissue structure. Yet, as we have seen, plants are susceptible to cancer-like growths (fasciations). These are not invasive but have all of the characteristics of cellular cheating as it relates to cancer: too much proliferation, a lack of appropriate cell death, monopolization of resources, a breakdown of the division of labor (disrupted flowering patterns), and destruction of the shared environment (for example, a greater likelihood of tissue death that can make the whole plant more susceptible to infection). And invasive growth can also happen sometimes in plants. One of the surprises we discovered in our review of cancer across the tree of life was a paper that reported an invasive growth in a plant—an invading front of cells breaking through existing tissues. This invasive growth satisfies even the more restrictive and traditional definition of cancer—which suggests that by all measures and definitions, plants can get cancer.

One important caveat to our first research project on cancer across the tree of life is that we looked only at published reports of cancer. This work helped us see the big picture regarding how cancer affects all branches of multicellularity in the tree of life. But it was one step of many. The next steps involve looking systematically at as many data as we can possibly collect about cancer across life.

I am now working with several members of the original Wiko working group—many of whom are now part of the Arizona Cancer Evolution Center—to achieve this goal, analyzing cancer rates across species. The cancer biologist and evolutionary biologist Amy Boddy, at the University of California, Santa Barbara, leads this project. Boddy is organizing a massive effort to create a comprehensive database of cancer records, bringing together data from zoos, veterinary clinics, and other sources. This database comprises approximately one hundred seventy thousand animal records, from thirteen thousand species. So far, we haven't found any animals in this database that are completely cancer resistant—every species with at least fifty animal records has at least one report of a neoplasm. The species with the highest cancer rates in the database at the time of this writing are ferrets, hedgehogs, and guinea pigs. Other animals with very high cancer rates include cheetahs and Tasmanian devils (even excluding cases of the Tasmanian devil transmissible facial tumors).

In addition to our work with this database of cancer risk across species, we are looking at placozoa and sponges—"simple" forms of life that we have found to be apparently cancer resistant in our review of cancer across life. (Placozoa and sponges are not part of the comparative oncology database we discussed above since they are typically not treated by veterinarians or in zoos, where those data come from.) My colleague and collaborator Angelo Fortunato is in charge of the project focusing on these ancient multicellular life forms and investigating their ability to resist cancer. Fortunato holds two doctorate degrees, in evolutionary biology and cancer biology, which gives him a unique set of skills and background for looking at the evolution of cancer suppression mechanisms. By studying the mechanisms of cancer resistance in these species, we can gain a better understanding of how cancer suppression originally evolved, and we may even uncover novel mechanisms of cancer resistance. We can also learn more about human disease and how we can better treat or prevent cancer in humans.

Fortunato has focused his efforts on a few species that seem to not get cancer (we found no reports of cancer in these species in our original review of the literature). One of the first species that Fortunato brought into the lab was a type of sea sponge, *Tethya wilhema*, an organism that is little more than a collection of largely undifferentiated cells with pores and channels that allow water and nutrients to flow through them. Fortunato has found that the sponges appear to be extremely cancer resistant; they can tolerate extremely high levels of radiation (which induces DNA damage) without having any apparent cancer-like growths. When he watches how these sponges respond to radiation, he has noticed that sometimes they shrink for a few days and then bounce back to their previous size, and there are no obvious odd growths or discoloration he can detect that might indicate cancer. Fortunato is now using molecular techniques to try to uncover the mechanisms that are responsible for this apparent resilience to DNA damage.

He is also looking at cancer resistance in placozoa (*Trichoplax adhaerens*), an organism that technically qualifies as an animal but is basically a bag of cells with a layer of cells on the outside that help it move. When Fortunato exposes placozoa to radiation, he sometimes observes darkened areas (which may be cancers) growing inside them. Sometimes these darkened areas move to the outside edge of the organism and then seem to be extruded, or pinched off, leaving the remaining cells in the placozoa free of these darker cells (figure 5.3).

This may be a cancer suppression mechanism that organisms without complex tissues and organ systems, like placozoa, use—essentially ditching potentially problematic cells. It might seem like a strategy that could only work for simple organisms, but upon further consideration, cell extrusion looks like it might be a viable strategy

FIGURE 5.3 After being exposed to radiation, placozoa sometimes develop darkened areas. These dark areas appear to get moved to the periphery and exuded, or pinched off, leaving the remaining cells in the placozoa free of these darker cells. It is possible that this cell extrusion is a cancer suppression mechanism. Placozoa exposed to X-rays at 160 Gy. Image captured at magnification 150X using brightfield lighting. Photo courtesy of Angelo Fortunato.

on the scale of tissues, even in large organisms such as humans. In the human colon, for example, cells that proliferate too much can get extruded by their cellular neighbors. The cells in the neighboring areas can create a ring of actomyosin (what muscles are made of) that literally squeezes out the problematic cells. A similar phenomenon has been observed in fruit flies (*Drosophila*) as well: normal cells can produce filamin and vimentin that create long armlike protrusions that expel mutated cells. This mechanism only works, however, if the cells around the mutated cells are normal, pointing to the importance of the tumor microenvironment in this process. The extrusion of mutated cells could serve to protect the body from the potential threat of cancer that damaged cells might otherwise pose. This extrusion mechanism for getting rid of mutated cells appears to not be limited to just placozoa.

Fortunato's work provides us with a glimpse into the ways in which simple organisms may have evolved to protect themselves against cellular cheaters, and it should encourage all of us to ask questions and expand our frame of reference—looking across the entire tree of life to better understand how we have evolved to suppress cancer.

More Cells, More Cancer?

In the last section I explored cancer suppression in small and relatively simple forms of life, but how about larger and more complex organisms—such as humans and elephants? How do large and complex forms of life keep cancer under control long enough to successfully reproduce?

Cell proliferation is necessary for being a multicellular organism, but it can also increase our susceptibility to cancer because mutations can arise anytime a cell divides. The larger an organism is, the more cell divisions it takes to get to that size, and the more cell divisions it requires to maintain that size (since tissues need to be renewed on an ongoing basis). In addition, the larger an organism, the more cells are around at any given moment that can mutate. In fact, if we look at cancer incidence within species, individuals that are larger have a greater risk of cancer. For example, larger breeds of dogs (heavier than about 20 kilograms, or 44 pounds) have a greater risk of cancer than dogs from smaller breeds and, similarly, taller humans have a greater risk of cancer than shorter humans—with about a 10 percent increased risk of cancer for every 10 centimeters (about 3.9 inches) of height. Yet this same pattern of greater cancer risk with larger size does not hold if we look across species.

At the opening of this chapter, we saw that elephants have one hundred times more cells than humans—yet they don't they get one hundred times more cancer than we do. Elephants are surprisingly cancer resistant for their size and longevity. In fact, they have lower incidences of cancer than many organisms that are much smaller, humans included. Mice, on the other hand, get much more cancer than we do despite the fact that they are much smaller than we are.

This paradox holds for longevity as well: the longer the lifespan, the more opportunities for cancer to arise as cells divide and are exposed to potential mutagens. Yet cancer rates do not correlate with longevity if we look at cancer risk across species.

This pattern—that cancer risk does not correlate with body size and longevity—is known as Peto's Paradox. It was pointed out by Sir Richard Peto, a statistical epidemiologist at Oxford, in the 1970s. He noted that—on a cell-to-cell comparison—human cells must be more cancer resistant than mouse cells; otherwise, we would succumb to cancer at an early age. Research that my colleagues and I have conducted over the past several years has confirmed this pattern: species with longer lifespans and larger sizes do not have higher cancer incidences than species that are smaller and shorter-lived.

Making Good Life History Decisions

All of us walk the tightrope in life, balancing cellular freedom and control. Too much cellular freedom and we increase our risk of cancer. Too much cellular control and we risk stagnation and evolutionary failure. It is no different for other multicellular organisms. Every organism must find the right balance between allowing cells to do what is necessary for the organism to stay alive and reproductively viable, while keeping cells under enough control so they don't become cancerous.

Not all organisms find the same balance. Some organisms, such as mice, spend their short lives leaning to the left on the tightrope, allowing cellular chaos to dominate until they get picked off by a predator. Other organisms, such as elephants, spend their lives angling to the right, leaning in on suppressing cancer so that they can live for a long time and reproduce later in life.

Elephants are playing the long game: they procreate later, and they don't have any natural predators, so they invest in cancer suppression in order to live long enough for that strategy to pay off. It's as if they have to traverse a much longer tightrope to reach the point where they can get a reproductive payoff. Not only do they lean a bit to the right in order to make it more likely they will achieve this reproductive payoff in the end, they also have to be better at the balancing act as a whole if they are going to make it as far as they need to for successful reproduction.

Some of the forces that lean organisms toward chaotic cellular freedom come from the inside of the organism. Other forces that lean the organism toward greater cancer risk come from outside of the organism, like DNA damage from exposure to solar radiation or chemical mutagens. There are also some influences that can come from experience—a wound is a good example. Wounding can increase the risk of cancer by activating genes that essentially increase the tolerance for chaos at the site of the wound.

As populations of organisms evolve over many generations of evolutionary time, other forces affect this balance between cellular freedom and control: forces such as high extrinsic mortality (the likelihood that you will die for external reasons, such as from predation), and intense sexual selection (when reproduction is strongly affected by your ability to attract the opposite sex and compete with the same sex). These evolutionary pressures can actually select for organisms that have a more left-leaning strategy, because the gains in cancer suppression aren't worth the trade-offs if you don't live long enough to reap those benefits (or if you must give up too many reproductive opportunities in exchange for a lower risk of cancer).

In evolutionary biology, these trade-offs are called life history trade-offs, because they affect the ways that organisms invest in various "goals" (like growth, reproduction, and survival) over the course of their lives. The basic idea of life history theory is this: Organisms have limited resources (like time and energy) to allocate toward various goals that could ultimately increase their reproductive success. Spending more on one thing leaves less time to spend on the other.

This is similar to our tightrope metaphor, but it expands to many dimensions. Growth trades off with reproduction, which trades off with survival, which trades off with growth, and so on. And we can break each of these goals into subgoals, many of which can trade off with each other too. One way of simplifying this concept is to think about the trade-offs mostly in one dimension—time—and, in particular, the timing of reproduction. Organisms that invest a lot in being viable early in life (growing quickly and creating as many offspring as early as possible) are fast life history organisms. Organisms that invest in long-term survival and viability but grow more slowly, delay their reproduction, and have fewer offspring are slow life history organisms. (Neither of these strategies is inherently better than the other; the best strategy depends on the ecology of the organism—and, in particular, the kinds of threats and opportunities they face.)

All else being equal, we should expect slow life history organisms like elephants to be less susceptible to cancer because they invest more in long-term survival rather than short-term reproduction. One aspect of long-term survival is somatic maintenance, which refers to anything an organism does to maintain its body, from wound healing to fighting infection to repairing broken DNA. Suppressing cancer is a part of somatic maintenance. Keeping the body free of cancer by detecting mutations and suppressing cellular cheating is one very important way of extending your lifespan if you are a multicellular organism with a slow life history strategy.

As we've seen, effective cancer suppression often comes at a cost. Doing too good of a job suppressing cancer can affect other traits relevant to fitness. This is one of the reasons organisms haven't evolved to suppress cancer entirely, despite millions of years of evolving with cancer. Too much cancer suppression can have a negative impact on life, as we've also seen.

Artificial selection in agriculture provides a unique window into these trade-offs between fitness relevant traits and cancer risk. We selectively breed animals for traits such as egg laying and milk production, and this intense artificial selection can sometimes yield surprising results that help us to understand trade-offs underlying certain traits. This is the case for hens, which have been bred for egg production. In addition to producing more eggs, they also have high rates of ovarian cancer, likely because they have been selected to have more permissive cell proliferation in and around their ovaries.

The rapid growth of seasonal antlers in deer is another example of the delicate balance of cancer suppression. In the winter, deer shed their antlers. Then, during the spring and summer, the antlers grow back rapidly in preparation for the breeding season in the fall.

FIGURE 5.4 Antleromas are bony masses that occur when antlers develop abnormally. Susceptibility to antleromas is partially a result of the rapid proliferation that is required to grow antlers each breeding season. Antlers are an example of a sexually selected trait that is associated with an increased risk of cancer.

The males that grow the largest antlers have a breeding advantage over other males. But this ability also leaves male deer vulnerable to strange cancer-like growths on their antlers, called antleromas (figure 5.4). The extremely fast growth of these antlers requires both rapid cell proliferation and tight control over growth so it doesn't get out of control. Many signs point to an association between cancer pathways and the ability to quickly grow antlers. Even normal antlers without antleromas have gene expression patterns more indicative of bone cancer (osteosarcoma) than normal bone. In addition, tumor promoter genes are expressed in antlers, and genetic sequencing has revealed that genes associated with cancer (proto-oncogenes) have been under positive selection in the ancestors of cervids (deer). These antlers are an example of a sexually selected trait (because females are more likely to mate with males with larger antlers) that increases susceptibility to cancer.

Another sexually selected trait that can increase cancer susceptibility within a species is large body size. In many species, females prefer to mate with larger males. One example is a freshwater fish called the southern platyfish (also known as the moonfish, or *Xiphophorus maculatus*). Some males of this species are significantly larger than others. These larger males are called X-mark males because they typically have a large black spot on their abdomens. This black spot is a melanoma. The same gene that is responsible for large size also makes these fish more susceptible to skin cancer.

Being a large organism requires more cell proliferation—both to get to that size and to maintain it. And this means a greater risk of cancer. But this pattern falls apart when we look at relationships among species. Elephants and other slow life history organisms have some evolutionary tricks up their sleeves that allow them to be both big and cancer resistant.

Elephants boast extra copies of the tumor suppressor gene *TP53*, which contributes to their low rates of cancer. (We have just two— one from our mother and one from our father.) As we have seen, *TP53* helps keep cellular proliferation under control and induces programmed cell death when cells are too damaged to be repaired. It acts like the cheater detector of the genome, monitoring cells for aberrant behavior and responding accordingly.

TP53 is one of many tumor suppressor genes, but it's one of the most important: it helps maintain cells in a healthy state by detecting problematic conditions like DNA damage. If it detects damage, it halts the cell cycle until the problem can be fixed. And if the problem cannot be fixed, *TP53* initiates cellular suicide, starting a signaling cascade that ends in apoptosis. With their extra copies of *TP53*, elephants get an extra dose of all this cancer suppression functionality; they are especially sensitive to DNA damage, so their cells self-destruct more readily if damage occurs.

The evolutionary cancer biologist Carlo Maley (who is also my collaborator and husband), along with his student Aleah Caulin, discovered these extra copies of *TP53* in the elephant genome and suspected that they may be playing an important role in the animals' low rates of cancer. Maley's findings caught the attention of Joshua Schiffman, the pediatric oncologist whose experience with his dog moved him to start studying similarities in cancer between humans and dogs. Schiffman had been irradiating cells and measuring apoptosis rates to understand the hereditary condition Li-Fraumeni syndrome, which I discussed earlier in this chapter. In Li-Fraumeni, babies are born with only one copy of the tumor suppressor gene *TP53*, rather than the normal two copies (one from Mom and one from Dad). Children with Li-Fraumeni have nearly a 100 percent chance of getting cancer in their lifetimes, and many get multiple cancers, sometimes starting in early childhood. It is a tragic and heritable condition: sometimes whole families will be plagued by it.

Schiffman found that when he irradiated cells from the blood of people with Li-Fraumeni syndrome, the cells responded to the radiation in an unusual way. Rather than undergo cell death when the DNA gets damaged, as a normal cell should, these cells stayed alive. This leads to more cell survival, but ultimately makes the whole body more susceptible to cancer. In Li-Fraumeni, cells stay alive when they have high levels of DNA damage because of a faulty copy of *TP53*, and these mutant cells can then threaten the life of the patient by increasing the chance of cancer.

Maley and Schiffman decided to team up and examine the DNA damage response in elephants' cells to see if these many copies of *TP53* were helping to protect elephants from potentially cancer-causing cells. They enlisted the help of Lisa Abegglen, a molecular pathologist and cancer biologist at the Huntsman Cancer Institute. When Abegglen and others on the research team irradiated the cultured cells from elephant blood, they found that the rates of apoptosis were sky high: the response of the elephant cells to radiation was to self-destruct. This hair trigger for cellular suicide helps to protect elephants from mutant cells that could give rise to cancer.

When researchers culture elephant cells in a Petri dish and then irradiate them, this activates the gene *TP53*, which then produces the protein p53, which subsequently induces cell death in cells that are highly mutated. In the language of our tightrope balancing act, the production of this gene product, p53, leans the elephant toward the right, increasing control over cellular behavior. When *TP53* gets activated (by radiation damage, for example), it produces p53, which goes into the bucket on the right, helping the elephant manage the increased risk of cancer that can come from exposures like radiation.

The team combined Maley's analyses showing that elephants have forty copies of *TP53* with the data from Schiffman's lab showing that elephant cells readily self-destruct in response to radiation. Their work combining computational biology and genomics with in vitro lab studies of cellular responses to DNA damage is a great example of the kind of innovation that can come from interdisciplinary teams working together to solve a long-standing puzzle like Peto's Paradox.

Other teams have replicated Maley and Schiffman's results and found further evidence suggesting that elephants solved Peto's Paradox by having multiple copies of *TP53*. In fact, the evolutionary biologist Vincent Lynch at the University of Chicago separately made the same discovery that elephants have multiple copies of *TP53*. By looking at DNA from woolly mammoths and other extinct relatives of elephants, Lynch and his team reconstructed the evolutionary changes in the number of *TP53* genes over time. They found that over the course of organismal evolutionary time the number of copies of *TP53* increased when body size increased. This finding suggests that increasing body size may have encouraged the evolution of more intense cancer suppression in the form of more copies of *TP53*.

Elephants are not the only organism that evolved cancer suppression systems to enable larger body size. Marc Tollis, an evolutionary biologist at Northern Arizona University (and also a member of our research team), discovered that humpback whales have duplications of apoptosis genes; and have had positive selection on genes responsible for cell cycle control, cell signaling, and cell proliferation compared to their smaller cetacean cousins (including the sperm whale, bottlenose dolphin, and orca).

Regulated and under Control

The balancing act between cellular freedom and cellular control is a dynamic, lifelong process. Genes like *TP53* are not always expressing proteins—if they were, this would shift the balance on our tightrope too far to the right, which would come with its own trade-offs (like premature aging or low fertility). Elephants are not just leaning more to the right—they are also balancing more carefully and actively than small organisms like mice. Being large and long-lived requires both better cancer suppression and more careful regulation of that cancer suppression to keep the organism balanced on this tightrope over the lifespan. It's not just a matter of producing more gene products that increase the levels of cellular control, it's also a matter of producing those gene products at the correct times and in the appropriate amounts so that they balance out the gene products that would otherwise lean the organism toward cellular chaos.

How do organisms regulate the regulators? One way is by creating genetic networks (connections between genes that allow them to be influenced by one another's states) that extend from the genes that promote cellular freedom to those that promote cellular control. By monitoring—and influencing—the production of gene products, these networks can help the organism maintain this delicate balance longer (as we saw earlier with the signal detection function of *TP53*).

The genes that are mostly responsible for tipping the balance toward cellular freedom (promoting cell replication) are the most ancient genes, the ones that arose during unicellularity. The genes that are mostly responsible for tipping the balance toward more cellular control evolved at the transition to multicellularity. Many of these genes, sometimes called caretaker genes, help enforce cellular cooperation to make the multicellular organism viable. But there is yet another category of genes: those genes that live between the "unicellular" freedom-promoting genes and the "multicellular" control-promoting genes. These genes, called gatekeeper genes, help keep the whole system in balance, dynamically responding to changes and sending signals to both sides to adjust as needed.

The gatekeeper genes that exist at the interface between "unicellular" and "multicellular" genes are the most evolutionarily recent. They make it possible for large, long-lived forms of life—like humans and elephants—to balance and negotiate the conflicting needs for cellular freedom and cellular control throughout their lifespans. These genes can help organisms dynamically manage the ever-changing forces that would otherwise destabilize this challenging tightrope act.

Of Devils and Dogs

Keeping cancer from evolving inside the body is a lifelong challenge. But it is not the only cancer-related challenge that organisms face. Since the beginning of multicellularity, life had to deal with the possibility of cancer evolving within—as well as the possibility of cancer invading from the outside. Many of the reports of transmissible cancers are recent—from the past decade or so—but transmissible cancer is as old as multicellularity itself.

What we now refer to as transmissible cancer was a major problem for the earliest forms of multicellular life. The first multicellular organisms were essentially collections of cells that were cooperating in order to survive and reproduce better as a group, rather than as individual cells. Early in the evolution of multicellularity, some cells specialized in trying to invade and take advantage of these cooperative cellular societies rather than creating and maintaining a cellular society of their own. Some cells specialized in invading the germ line, co-opting the whole multicellular society for their own reproduction, a process called germ line parasitism. Other cells specialized in invading a stem cell niche, co-opting this cell renewal system to make copies of themselves, a process called stem cell parasitism. In order for multicellular life to be viable, early organisms had to evolve ways of keeping these invaders out. One of the most important adaptations for keeping these invaders at bay is the immune system.

The risk of being invaded by germ line and stem cell parasites was one of the first selection pressures on the evolution of the earliest immune systems. Since the initial evolution of multicellularity, immune systems have evolved to be much more complex. Our immune system includes both the innate immune system, which is a rapid and general response to threats by cells including natural killer cells, and the adaptive immune system, which is a longer-term response to specific threats that works through leveraging somatic evolution, as we saw in the last chapter. Our immune system also includes the skin, which helps protect us from external threats.

When something goes wrong in the immune system—whether it is a breach of the skin barrier, the hijacking of immune cell replication, or an interference with the ability of the immune cells to identify threats—this increases the risk that cancer cells will successfully jump from one organism to another.

About ten thousand years ago, the Alaskan malamute, a beautiful sled dog with a double coat of fur that protects it in cold conditions, gave rise to a new species of dog that was quite unusual. This new species likely came about simply due to mating between a male and a female—but it resulted in a novel species, and one that looked nothing like its ancestor. In fact, it looked nothing like a dog at all. It was a single-celled species of dog that made its living as a sexually transmitted parasite—a sexually transmitted cancer.

This strange species is known as canine transmissible venereal tumor (CTVT). It is considered a species of dog though it has no wagging tail, no floppy ears, no friendly eyes—it is just an unpleasant-looking mass of cells growing on the genitals of the dogs afflicted with it. Since its beginning it has spread via sexual contact, sniffing, and licking to dogs on every continent in the world except Antarctica (this spread most likely being made possible through the movement of human populations through shipping, etc.). Some researchers believe that CTVT was so successful that it may have even led to the extinction of the first dogs in North America. This dog transmissible cancer owes its origins and success to the behavior of the host species—in particular their sexual behavior. After coitus, dogs often become "tied" together due to the expansion of the penis in the female's genital tract. When the dogs try to separate, the genital area can be injured, breaching the first line of defense of the immune system—the skin. These wounds can create an environment that is more conducive to the growth of the transmissible cancer cells. As we have seen in previous chapters, when a wound is healing, tissues become more tolerant of cell proliferation and cell movement. This may allow cancer cells—in this case, transmissible cancer cells—to fly under the radar of the dog's immune system, proliferating inside the wound and perhaps even taking advantage of the growth factors that are part of that healing response.

Every transmissible cancer cell from every dog with CTVT originated from a single dog about ten thousand years ago. This makes it the oldest known somatic cell line (orders of magnitude older than the oldest somatic cell line humans have cultivated, HeLa cells, which were derived from cervical cancer cells taken from Henrietta Lacks in 1951). CTVT is also the only known unicellular species of dog, and the only obligately parasitic species of dog (though some might argue that certain breeds—like the French bulldog, which is dependent on humans to complete its reproductive cycle through surgical C-section—are also obligate parasites). The CTVT cells are much like other unicellular infectious agents, surviving long after their hosts have died and continuing to successfully transmit to other members of the host population. The unicellular dog is not the only species that has evolved from multicellular ancestors to become a unicellular infectious agent: enter the Tasmanian devil.

Among Bugs Bunny's crew of dim-witted stalkers is the omnivorous, unpredictable, and very bitey Tasmanian devil, Taz. This wild, erratic character bites through anything, creating a destructive high-speed vortex of teeth and claws when it gets excited or agitated. The real Tasmanian devil doesn't look much like Taz, but Looney Tunes did get some things right. Tasmanian devils are carnivores—the largest living marsupial carnivores in the world—and their propensity to bite one another has contributed to the spread of a deadly, contagious cancer that has landed Tasmanian devils on the Endangered Species List.

In 1996, scientists observed a strange facial tumor in Tasmanian devils in the northeast corner of Tasmania, an island off of Australia's southeast coast. This condition, which came to be called devil facial tumor disease (DFTD), grows around the mouth and other facial areas (figure 5.5); and when the devils engage in their notoriously aggressive behavior, pieces of the tumor can flake off and infect the wounds of their rivals. As with the dog transmissible cancer, the cells from DFTD tumors can grow in the wounds that happen during these injury-inducing social interactions. (This is also very similar to what happens in a surgical allograft: the tissues from one individual are transplanted to another individual, where they can take hold and

FIGURE 5.5 Tasmanian devils are afflicted by a facial tumor that can transmit cancer cells through aggressive encounters. During fights, Tasmanian devils often bite one another's faces, which can allow cancer cells to flake off the face of one individual and then land in the wounds of the other individual. If these transmissible cells land in a new wound that is filled with growth factors and inflammatory signals, it can often grow in the new individual like an allograft. Transmissible cancers also are known to occur among dogs and clams. They are extremely rare in humans.

survive. Like an organ transplant or skin graft, the DFTD cancer cells can grow and thrive in a new host.)

Devil facial tumor disease is deadly, and it has been spreading and threatening the Tasmanian devil population in a big way. DFTD usually kills its host within six to twelve months, but in that time, the host has countless opportunities to transmit the facial tumor to other devils, as fighting and biting are part of many aspects of Tasmanian devil social behavior—from aggressive mating to male-male competition.

Mating and fighting provide opportunities for DFTD cancer cells to flake off and land in open wounds. As with the dog transmissible cancer that appears to thrive in the mating-related wounds around dogs' genitals, the devils' transmissible cancer cells thrive in the wounds on devils' faces. The breach of the skin barrier creates an initial vulnerability. Then, inflammation and cell proliferation that are normal parts of the wound healing response create an environment that is particularly conducive to the growth of transmissible cancer cells.

There are many parallels between the devils' and the dogs' transmissible cancers. But unlike the dog transmissible cancer, the Tasmanian devil facial tumor does not have one single origin. It has two: one from a male and one from a female. The first Tasmanian devil facial tumor discovered, DFTD1, has two X-chromosomes, indicating that it originated from a female. The second one discovered, DFTD2, has a Y-chromosome, indicating that it originated from a male.

This suggests that transmissible cancer in Tasmanian devils may not be so rare, and perhaps that the evolution of transmissible cancers in general might not be the exceptionally rare event that it was previously thought to be. Elizabeth Murchison, one of the researchers studying DFTD, told me that when she and her colleagues began studying transmissible cancer in Tasmanian devils, they thought transmissible cancer was extremely rare in nature. But when they discovered in 2016 that there was a second separate origin of transmissible cancer in devils, they were forced to take a fresh look at their assumptions about contagious cancers. Contagious cancers may not be as rare and bizarre as we've thought.

One of the reasons that transmissible cancers exist at all is that they can fly under the radar of the immune system. When populations of organisms have low genetic diversity—like Tasmanian devils and dogs do—then cancer cells that evolve to evade the immune system in one individual may already be able to evade the (genetically similar) immune system of a new individual in the population.

Transmissible cancers like those in dogs and devils grow like an allograft (a transplanted tissue) in the open wounds that arise from fighting and mating. If the immune system is functioning properly, it will typically identify the foreign tissue and reject the allograft. This is why organ transplants require tissue matches (to ensure compatibility between donor and recipient) as well as immune-suppressing drugs.

One reason that transmissible cancers have spread in dogs and Tasmanian devils may be that both dogs and devils are fairly genetically homogenous. This essentially makes the transmissible cancer cells a good "tissue match" for new potential hosts, facilitating the growth of this cancer allograft without the immune system of the new host rejecting them. Both dogs and devils have gone through genetic bottlenecks, periods in history when the genetic diversity of the population went down. In dogs, this was largely the result of selective breeding by humans to create the breeds we have today—many of which are highly inbred. In Tasmanian devils, their genetic homogeneity is a result of population bottlenecks that reduced their genetic diversity—largely because of campaigns to eliminate them on the part of European settlers who arrived in Tasmania in the nineteenth century. From the perspective of a contagious cancer cell, it can be easier to jump from one host to another in a more genetically homogenous population, since the new host might be very similar to the previous host, immunologically speaking.

Genetic homogeneity is not the sole reason why Tasmanian devil cancer cells can hide from the immune system. DFTD cells can make themselves largely "invisible" to the immune system by downregulating their expression of major histocompatibility complex (MHC), the molecules on the outside of the cell that display pieces of the proteins used inside the cell, allowing the immune system to tell self from nonself. This is a common strategy that human cancers also use to evade the immune system. By lowering their MHC levels, DFTD cells can better proliferate and migrate around the body without prompting an immune response.

Like the devil transmissible cancer, CTVT also messes with the self/nonself signaling systems on the outside of the cell. In CTVT, the MHC "tags" on the outside of the cell that reveal its identity are downregulated, making it easier for these cells to go undetected. However, in the case of the canine transmissible cancer cells, the MHC tags may be hidden initially, but express themselves later in progression for reasons that are unknown. Sometimes, the canine transmissible cancer can regress—and this regression is associated with an increase in MHC expression and the presence of immune cells at the site of the tumor. These similarities and differences between the devil tumor and dog tumor suggest that the immune system may be limiting transmissible cancer in mammals more generally.

The invasion of cellular cheaters from the outside of the organism may be more common than we think. Even today, cells such as the transmissible cancer cells from Tasmanian devils and dogs continue to break free of the multicellular organisms that they evolved to cooperate within. These cells set off on a journey beyond the organism from whence they came, usually to meet their demise in the hands of a hostile external world, but sometimes they succeed, adopting a transmissible lifestyle, colonizing new individuals, and spreading to a new host.

Transmissible cancer cells are not just a problem for land-dwelling forms of life—they are a problem for water-dwelling life as well. In fact, they may be a much bigger problem for water-dwelling organisms like clams that routinely come in contact with potentially cancerous cells floating through the water. Bivalves, a class of aquatic invertebrates that includes clams, mussels, scallops, and oysters, can be found in both saltwater and freshwater. They are filter feeders, which means that they need no head, mouth, or means of movement to capture their prey. They simply sit on rocks or bury themselves in underwater sediment where they can just let the world literally wash over them as they extract the nutrients they need to survive.

But this filter feeder approach—letting everything into the body—introduces the possibility that free-floating cancerous cells will enter the fold. Bivalves often live in colonies with genetically similar individuals nearby, and so cancerous cells shed by one bivalve may find their way into the filter of another one nearby.

The bivalve immune system involves a number of different layers of defense, from the outside shell, to mucosal surfaces, to the activation of blood cells called hemocytes that play a role in detecting and responding to potential infection. Ideally, these hemocytes help protect bivalves from infection, but unfortunately they can also leave them vulnerable to fatal leukemia-like cancers, which occur when hemocytes proliferate too much. These leukemia-like cancers have been found in at least five different bivalve species to date. Michael Metzger, a marine biologist at the Pacific Northwest Research Institute (who was working in the lab of marine biologist Steve Goff at the time) suspected that some of these leukemia-like cancers in bivalves were the result of transmission of cancer cells from one individual to another. Metzger looked for signatures of transmissible cancer cells in five bivalve species known to have these leukemia-like cancers. In every case, the cancer was attributable to bivalve contagious cancer cells. Metzger even found that one of these cases was a cancer transmitted *across* species—from a pullet carpet shell to a golden carpet shell clam. He suspects that transmissible cancer may be responsible for bivalve leukemia-like cancers more generally.

The prevalence of transmissible cancer in clams is also challenging the received wisdom that transmissible cancers are rare in nature and in the history of life on this planet. Metzger suspects, as do I, that transmissible cancers have been a selective pressure on organisms since the beginnings of multicellular life. Perhaps we don't see many instances of transmissible cancers because multicellular life has been under very strong selective pressures to protect ourselves from it—through the development of a variety of immune barriers, for example. But when the immune system is compromised, contagious cancers may be able to get a foothold.

We know that bivalves are susceptible to contagious cancer because of their water-dwelling existence and method of feeding. In addition, as invertebrates, bivalves have immune systems that are quite different from the immune systems of vertebrates like us. They do not have MHC molecules on the outsides of their cells like we do, and we are just beginning to understand how their immune systems protect them from external threats.

Metzger suggests that bivalves may use a self/nonself recognition system like those of sea squirts, called fusion/histocompatibility (Fu/HC) systems. Fu/HC systems help protect sea squirts from stem cell parasitism, which can happen when unrelated sea squirts fuse together and cells from one sea squirt begin proliferating in the stem cell niches of the other.

The problem of contagious cancer in bivalves may not be so different from the problems that early multicellular life had to solve in protecting itself from invading cells. As we saw earlier, stem cell parasitism and the related phenomenon of germ line parasitism were problems during the evolution of multicellularity.

Cross-species cancer is not limited to the clams that Metzger has studied. In 2013, an HIV-positive man went to the hospital with symptoms of a fever, cough, and weight loss. A biopsy of the patient's lymph nodes revealed strange cells that looked like cancer but were much smaller than human cells. This led the doctors to wonder if perhaps it wasn't cancer, but rather an infection from a unicellular eukaryote, like slime mold.

The unicellular appearance of the cells and absence of tissue architecture put them off the trail of what these growths really were. Genetic analysis showed that they were actually tapeworm cells, growing as a cancer in the tissue of the patient. Reports of this condition claim that it is extremely rare, though there are actually several other reports of tapeworm cells growing in humans in this way. And they have something important in common: every one of them occurred in an individual with a compromised immune system. Three of the four reports of this tapeworm cancer noted that the patient was HIV-positive, and the fourth individual had a compromised immune system due to Hodgkin's lymphoma. We don't yet know exactly how this tapeworm cancer arises within the host—whether the tapeworm developed cancer within the host, which then grew out into the host tissues, or whether another process was responsible—but what is clear is that suppression of the immune system is the common denominator in all of these cases. Most likely, this cross-species tapeworm-human transmissible cancer couldn't have taken hold in these individuals if the host's immune systems had been functioning properly.

Transmissible Cancer (Almost) Never Happens in Humans

Transmissible cancer may be more widespread than we thought, but luckily it is not a problem in most species, including humans. This could be because our immune system is simply better than the immune system of bivalves, or it could be because we are genetically diverse enough that transmissible cells can't take hold. (Or perhaps we are less likely to be exposed to potentially transmissible cells because we are not as violent or aggressive as Tasmanian devils—or at least that our violence and aggression rarely turns into face-biting. Though if you are worried about a zombie apocalypse, yes, you can now worry that it could create conditions favoring the evolution of transmissible cancer in humans. But in the zombie apocalypse, this would be the least of your worries.)

Organ transplants save thousands of lives every year. But sometimes organ recipients get more than what they bargained for. In some rare cases, organ recipients end up with cancer. Genetic tests confirm that the cancer actually came from the organ donor (rather than being a result of a new tumor arising in the recipient). This phenomenon was first recognized by the transplant surgeon Israel Penn. He noticed an apparent increase in cancer among transplant recipients and created a registry for transplanted tumors to facilitate the study of these transplanted cancers and develop screening protocols to reduce their likelihood.

These cases are extremely rare: a study of more than one hundred thousand donors found only eighteen cases, making the donor-related tumor rate for transplanted cadaveric organs incredibly small, at 0.017 percent (this was in a study of non–central nervous system (CNS) tumors; donor-related CNS tumors are also vanishingly rare; a study of hundreds of donors with CNS tumors found no examples of transmission). The chance that you will die while waiting for an organ to come through on the list is much, much higher than the chance of getting cancer from an organ donor—so the possibility of transmissible cancer is certainly not a reason to refuse an organ donation if you need one. The benefits of getting an organ transplant far outweigh the risk of getting a transmissible cancer from the procedure. And screening is constantly being improved and updated to better prevent the inadvertent transplanting of tumors from donors to recipients.

As with the dog and Tasmanian devil transmissible cancers, organ transplant transmitted cancer involves a breaching of typical immune defense. In organ transplants, the immune system is compromised. First, the skin is breached during the surgery to transplant the organ. Next, immunosuppressive drugs lower the body's defenses against foreign cells, making it less likely that the organ will be rejected, but simultaneously making it less likely that the immune system can properly detect and reject cancer cells that may be hitching a ride on the transplanted organ.

Occasionally, tumors have been inadvertently transplanted from patient to surgeon because of accidental injury during the surgery.

During an operation to remove a sarcoma, a surgeon accidentally cut his left hand. Five months later, a tumor was growing where this cut had occurred. Genetic testing showed that the tumor originated from the patient that the surgeon had operated on. In another case, a laboratory worker accidentally punctured herself with a needle containing colonic adenocarcinoma cells, leading to the growth of a tumor nodule. In both cases, the tumors were localized, and after they were removed, there was no evidence of recurrence. These two individuals who were accidentally transplanted with cancer had healthy immune systems, but were susceptible because the major immune barrier of the skin was compromised.

There are also some rare cases of cancer transmission in the womb. If we look over the last several decades, there are only about twenty-six reports of transmission of cancer from mother to fetus, including melanoma, leukemia, and lymphoma. This makes maternal-fetal transmission exceedingly rare given how many births there are and how many women have cancer while pregnant (cancer biologist Mel Greaves estimates the chance that a pregnant woman with cancer would transmit that cancer to her fetus to be around one in five hundred thousand). Some of these cases involved the loss of MHC molecules on the surface of the cells—as we saw with the case of the Tasmanian devil transmissible tumor, loss of MHC makes it easier for cells to evade the immune system. In addition to maternal-fetal transmission of cancer, there are also many reports of leukemia being transmitted between monozygotic twins in the womb. These twin-transmitted leukemias in the womb involve immunological invisibility (in which the immune system can't "see" invading cells because they look just like the cells of the organism), since monozygotic twins are genetically identical.

If you're worried about contagious cancer, don't be—these cases are exceptions to the rule. Contagious cancer in humans is extremely rare. Still, these curious cases give us insight into why contagious cancer is generally not a problem for us. When contagious cancer occurs in humans, it is associated with a disruption to the host immune system, like immunosuppression as a result of disease, medications, or the compromising of the skin.

Contagious cancer is not generally a problem for humans because we have particularly good mechanisms for detecting foreign cells and keeping them from getting out of control. Given our long evolutionary history of grappling with potential invaders from the outside—starting with germ line and stem cell parasitism in the origins of multicellularity—it is quite likely that some of our cancer suppression mechanisms, especially those involving our immune systems, may have evolved partially as contagious cancer suppression mechanisms—mechanisms to prevent, suppress, and respond to transmissible cancer threats. It is likely that multicellular life evolved not just to control cancer from within, but to prevent colonization by potentially contagious cells from other individuals.

Some of the mechanisms that protect us from contagious cancer are clear. But there is still an open question regarding whether protecting us from contagious cancer is part of what the immune system evolved to do: whether it is an evolved adaptation, or whether it is a by-product of other functions (like protecting us from pathogens). Did contagious cancer in our evolutionary history help shape what our immune system does to protect us, or it is just a side effect of the immune system's other abilities?

Our immune systems may indeed have evolved partly to protect us from contagious cancer. It may be that high levels of MHC diversity, for example, help protect vertebrates from transmissible cancers. A number of scholars who study transmissible cancer have proposed that our MHC diversity, one of the cornerstones of our vertebrate immune system, may have arisen because of the selective pressures that came from transmissible cancers. Although the idea that contagious cancers have shaped our immune systems is still speculation, there is no doubt that our immune systems do protect us from contagious cancers.

Another fascinating possibility that has been proposed is that sexual reproduction might have evolved in part to reduce the risk of contagious cancer. One of the prevailing theories for the evolution of sex is that it creates genetic diversity that makes offspring less vulnerable to transmission of infections. In other words, parents make offspring who are not identical to them because the offspring is therefore less likely to get bacterial and viral infections from the parent. We have already seen how genetic homogeneity contributed to the transmission and spread of contagious cancers in populations of dogs, Tasmanian devils, and even bivalves. Thus, the theory states that sexual reproduction can increase genetic heterogeneity in the population, decreasing the vulnerability of offspring to transmissible cancer. If sexual reproduction did evolve in part because it reduced the risk of contagious cancer, it is certainly ironic that dog infectious cancers are transmitted through sexual contact.

Transmissible cancer has been a problem since the origins of multicellularity; it evolved in the first place because the very first multicellular organisms could be parasitized by individual cells poised to take advantage of them—invading them, using their resources, and transmitting further. Today, transmissible cancers continue to invade multicellular bodies and exploit them to facilitate their own transmission. We don't know how many species in the history of life on earth have gone extinct because of transmissible cancers, but we have evidence that it has happened even in relatively recent history. Earlier in this chapter I discussed how transmissible cancer may be the reason for the extinction of the first dogs in North America. If extinction as a result of transmissible cancer has been even an occasional occurrence, it would have been an important selective pressure in the history of multicellular life.

6

The Hidden World of Cancer Cells

According to Joel Brown, cancer cells are a lot like squirrels. Brown is an ecologist and now also a cancer biologist, who is fond of squirrels and ecological analogies. Cancer cells, like squirrels, need resources to survive and, like squirrels, they face threats from the environment. All organisms need to find resources and avoid threats. No matter what kind of organism you are, whether you survive and thrive is a matter of how well you find resources and avoid threats.

Just like squirrels, cancer cells have to make a living in their environment. This means finding sources of food, protecting themselves from threats, and outrunning (or outsmarting) predators. Just like organisms evolving in the natural world, cancer cells that do a better job of accomplishing those goals will survive better and leave more descendants in the next generation.

In earlier chapters, I explored the benefits that can come from looking at cancer from the perspective of cancer cells: it is a way to better understand their strengths and weaknesses, a way to predict how they are likely to evolve, and—most important—a route to prevention and treatment. Taking cancer's perspective is also helpful when we look at the world in which cancer cells evolve. Our bodies are ecosystems in which cancer cells live, die, and change over time. From the perspective of a cancer cell, our bodies provide raw materials they need to proliferate, but also threaten them with immune destruction. Our tissues, our bloodstream, and even the signaling systems that our bodies use to share information can all be used by cancer cells to enhance their survival and help them replicate more quickly. Our organs are like different continents that can be colonized, our blood supply is like a nourishing river system, and our immune cells are like predators that must be outrun or avoided if the cancer cells are to survive.

In this chapter, I will look at cancer from this ecological perspective, and consider how cancer cells evolve, first to cheat by extracting resources from the multicellular body, and later to actually coordinate and cooperate with one another to better exploit the body. They evolve to signal for blood vessels, invade through membranes, and colonize new ecological environments in the body during metastasis. Cheating and cooperation both happen within this ecological context of the multicellular body. Cancer is an ecological problem in the body just as much as it is an evolutionary one.

The Making of the Tumor Microenvironment

Cancer cells live and evolve within a complex ecosystem composed of physical infrastructure (including collagen and enzymes that make up the extracellular matrix), other cells (both cancerous and normal), resources (from the blood and other cells), and threats (such as immune cells preying on them). This ecosystem (often referred to as the tumor microenvironment) influences how cancer cells evolve and behave. As cancer progresses, cancer cells change the tumor microenvironment by exhausting resources, building blood vessels, and hijacking normal "support cells" (for example, stromal cells) in nearby tissues. These changes in the tumor microenvironment can then alter cancer cell evolution and behavior, leading to a feedback loop between the ecology of the tumor and its evolution.

There are two main ways that changes in the tumor microenvironments can influence cancer progression: by changing how cells evolve inside the tumor and by changing what genes are expressed by the cancer cells. First, different microenvironments affect the evolutionary trajectory of cancer because they change the survival and reproduction prospects for any given cell. This can lead to evolutionary changes for the whole population of precancerous cells, often selecting for cells that are more cancer-like. Second, different environmental conditions can change the gene expression states of cells. These changes in gene expression affect cell behavior, allowing for different physiological capacities in different environments. For example, cells in environments with low oxygen will upregulate hypoxia-inducible factors (referred to as such because they are induced by hypoxic, or low-oxygen, conditions). These hypoxia-inducible factors can change how cells behave, leading them to become more motile, to signal for blood vessels, or alter their metabolism.

Some of the earliest work on the tumor microenvironment revealed that cancer cells can behave like normal cells if you put them into an environment with normal cells. Under normal microenvironmental conditions, the signals that these cancer cells get from their neighbors can keep them in an expression state that makes them behave like normal cells. Cancer is not just a matter of the genetic mutations that a cell has, but the environment that it finds itself in or creates for itself (for example, whether it has neighbors that suppress its cancerous behavior or promote it).

The centrality of the tumor microenvironment to keeping cancer at bay is also the main idea of the "tissue organization field theory" of cancer, which was originally proposed as an alternative to the somatic mutation theory (which posits that cancer arises from genetic mutations). But these two frameworks are not incompatible: genetic mutations and the tumor microenvironment interact throughout progression in ways that can either suppress or promote cancer.

Precancerous cells that gain a foothold in the body usually do so within a microenvironment conducive to their growth, sometimes called a tumor-promoting microenvironment. A conducive microenvironment might simply be an area near blood vessels or a region of tissue inside an organ that has high levels of hormones or other growth factors. These environments provide resources and factors that can potentially be used by cancer cells. Injury and tissue damage are other potential catalysts for tumor-promoting environments; they can both trigger a wound healing response that signals neighboring cells for rapid regrowth to cover a wound. As we have seen in earlier chapters, this healing response can create an environment in which cancer cells can thrive. Wound healing signaling is an example of a positive feedback loop between the ecology of a tumor and its evolutionary dynamics.

The tumor microenvironment changes as cancer progresses. Early on, cancer cells mostly just exploit the resources that are present in the tumor microenvironment. Later in cancer progression, cancer cells actually help bring new resources into the tumor by evolving the ability to signal for new blood vessels. These blood vessels can provide oxygen and nutrients that can fuel tumor growth. In addition, cancer cells can co-opt their neighboring stromal cells to provide growth and survival signals, constructing a niche conducive to their own proliferation.

Another important change in the tumor microenvironment that happens during cancer progression is the increase in immune cells. Tumors tend to attract vastly higher levels of immune cells than healthy tissues do. We have seen that immune cells help keep cancer cells at bay, but sometimes immune cells can get hijacked and actually help cancer cells. Chronic inflammation is one common characteristic of tumor microenvironments. Cancer cells evolve to signal to immune cells to produce growth factors, survival factors, and angiogenic factors for them, for example, by exploiting the signaling systems underlying the wound healing response, as I described earlier. Cancer cells can also recruit T-regulatory cells, the immune cells responsible for shutting down an immune response once a threat has been eliminated. In this case, the cancer cells make a nice safe niche for themselves by getting the T-regulatory cells to actually protect them from being killed by the immune system. It's as if cancer is manipulating these T-regulatory cells to tell the rest of the immune system, "Nothing to see here folks, move along," or "These are not the cells you're looking for."

All of these aspects of the tumor microenvironment can be understood in the context of ecology: the ecology of the body and the ecological interactions cancer cells have within it. And these ecological processes shape the way cancer cells evolve in the body and which genes the cancer cells end up expressing. There are numerous parallels between the ecological processes that happen in the natural world and those that happen in the cancer microenvironment, including niche construction, dispersal evolution, life history evolution, and even the tragedy of the commons and other social dilemmas. Understanding the changes to the body's ecosystem that happen in cancer, and the way that cancer cells evolve within that ecosystem, is essential for understanding how cancer cells evolve to cheat, but also how tumor cells cooperate with each other to better exploit the body.

What resources do cancer cells need to survive in the body? And how do they get the resources that they need? Cancer cells depend on resources delivered through the bloodstream, including oxygen and glucose. In addition, cancer cells require nitrogen and phosphorus to build nucleotides (the adenine, thymine, guanine, and cytosine that make up our DNA) and replicate their DNA—billions of new nucleotides must be synthesized for a cell to reproduce, leading to tremendous demand for nitrogen and phosphorus at tumor sites.

In addition to these basic resources, cancer cells also need growth and survival signals from their neighbors, at least early on in progression until cancer cells evolve to produce their own survival signals and growth factors. Cancer cells can evolve to co-opt the normal support cells (fibroblasts) in their environment to send resources; one way to do this is for the cancer cells to send out wound-healing signals to these support cells, which then prompts the support cells to send back growth and survival factors.

But our bodies are not passive vessels for cancer cells. Unlike squirrels that live in environments that—metaphorically speaking—don't care whether they are there or not, our bodies do "care" whether cancer cells are there. Our bodies go to great lengths to keep out cancer cells and limit their ability to survive and thrive within us. We have cancer suppression systems that help keep cancer cells in check. Many of these systems work by limiting the ability of cancer cells to create a favorable niche for themselves inside the body or expand in that niche.

Like organisms living in the natural world, cancer cells face hazards as well, many of which come from our evolved cancer suppression systems. One of the most important of these hazards is predation from the immune system. As we have seen, the immune system is a major component of our multitiered cancer suppression systems. Immune cells patrol the body for cells that are overproliferating, expressing mutant proteins, or expressing proteins from genes that they shouldn't be expressing. When immune cells find these clusters of abnormal cells, they produce factors that shut down proliferation, induce cell death, and block the recruitment of blood vessels to cut off the tumor from resources. In turn, cancer cells evolve to evade the immune system. This is very much like how prey animals evolve to evade predators. In fact, cancer cells evolve to evade the immune system through strategies like hiding (removing the markers on the outside of the cell that immune cells can identify) and camouflage (expressing genes that give them a more "normal" appearance to immune cells), just like organisms do in nature.

As cancer cells evolve to exploit the ecosystem of the body, they fuel their growth but they also fuel their own subsequent evolution— changing the environment as they go, often affecting the selection pressures on themselves and other cells around them. For example, by generating waste products like lactic acid, they can create an environment that favors cells with the ability to survive in high-acid environments. By using up all the resources in their local environments they can create conditions that favor cells that can disperse in order to find new environments to colonize, driving invasion and metastasis. After cells have exhausted local resources, they are under evolutionary pressure to hijack the resource delivery infrastructure of the multicellular body: our blood vessels. As we will see later in the chapter, cellular cheating can even generate selection pressures for cancer cells to cooperate with one another to signal to blood vessels, invade, and metastasize.

Inside the ecosystem of the body, cancer cells can evolve to make the environment around the tumor more conducive to their survival, for example, by hijacking blood vessels to increase the flow of resources or inducing stromal cells to provide more growth and survival factors in the milieu. But cancer cells also evolve to destroy the very environment that they depend on. As cancer cells exploit and colonize the tissues of the body, they present us with a paradox: on one hand, they destroy their local environments, exploiting resources and polluting the extracellular environment with lactic acid and other waste products. On the other hand, cancer cells build and cultivate local environments that protect and feed them, signaling for resources from blood vessels and hiding from the immune system. How is it possible that cancer cells are capable of both such dramatic destruction and such sophisticated creation?

Not all cancer cells are the same—some of them may rapidly exploit available resources while others signal for more. But both creative and destructive tendencies can be advantageous for cancer cells, depending on the context. Let's take a closer look at some of the ways that the ecological context can shape cancer cell strategies in the ecosystem of the body.

Cells have different strategies for making various trade-offs, just as organisms do. Some cancer cells will prioritize rapid growth and cell division, while others will prioritize survival. As we saw in chapter 5, these kinds of trade-offs are life history trade-offs, and organisms evolve to have different life history strategies in different environments. Cancer cells have life history strategies too. Some prioritize rapid replication, while others invest more in cellular survival. And like organisms, cancer cells evolve to have different life history strategies depending on the environments they are in.

Environments with stable resources and low hazards select for cancer cells with a slow life history strategy. Like the slow life history elephants I discussed in chapter 5, slow life history cells don't reproduce quickly and instead invest more in survival. On the other hand, environments with resources that fluctuate a lot have high levels of hazards, and so they will select for cancer cells with a fast life history strategy. Like fast life history mice, fast life history cells reproduce quickly and don't invest much in long-term survival. As we saw, the ecology of a tumor is typically filled with hazards. Tumors often have an erratic blood supply (since blood vessels often develop haphazardly, growing, branching, and collapsing at any time) and are infiltrated with immune cells that can prey on cancer cells. Both the erratic blood supply and immune cells can kill cancer cells, making the environment for cancer cells more hazardous and favoring the evolution of faster life history strategies.

Life history trade-offs help us understand how different tumor ecologies can select for different cell trade-offs. But these trade-offs might not emerge until later in progression. Early in progression, cancer cells evolve strategies to get around resource limitations, like altering their metabolism, signaling for more resources, and monopolizing the resources around them. Because cancer cells typically have access to a lot of resources early in tumor progression, they don't have to grapple so much with trade-offs (between cell proliferation and cell survival, for example). They can proliferate quickly while still having plenty of resources to allocate toward other "goals," like survival. But eventually—like any organism replicating out of control in an ecological system with limited resources—cancer cells find themselves in depleted environments where resources are no longer plentiful. When they run up against these resource limitations, trade-offs between proliferation and survival become more important.

Life history trade-offs are also likely to be very important during cancer treatment. Cancer therapy changes the ecology of the tumor and creates an environment that imposes trade-offs on cancer cells. During chemotherapy, cells can use efflux pumps (which are special molecular pumps) to detoxify themselves if the environment around them is filled with chemotherapy drugs. But these pumps need a lot of cellular resources to run; they are little machines that require energy to pump the toxins out. Approximately half of all the energy of the cell is required to operate these pumps. Cells that allocate resources to efflux pumps inevitably have fewer resources to allocate to cell division. We will see in the next chapter how trade-offs like these can be leveraged to design new approaches to treating cancer.

Squirrels and birds make nests, rabbits dig burrows, beavers make dams, and honeybees create hives. Like these organisms that transform the natural world to make it more favorable for them, cancer cells transform the internal world of our bodies to promote their own survival and reproduction. In ecology, niche construction is the process by which organisms alter their environment to make it more habitable, resource rich, and safe. Cancer cells are masters of niche construction: they shape their ecological niches through signaling for resources, protecting themselves from the immune system, and using several other strategies I will explore in this chapter.

In order to construct a niche for themselves, cancer cells must overcome many barriers of tissue architecture and growth regulation. One of the first steps involves invading the basement membrane, a membrane that forms a barrier between the outside of an organ and the internal cavity. Breaking through this barrier and others often requires cancer cells to cooperate with each other to produce factors (called matrix metalloproteinases) that break up the membrane. Getting through the body's membranes and other tissues also requires that cancer cells coordinate their electrical signaling. Cancer cells can also hijack normal support cells, called stromal cells, and induce them to start providing benefits to the cancer cells. These stromal cells can construct a cancer niche by producing growth factors, remodeling the tissue architecture (for example, producing collagen, which can make tumors feel like nodules in otherwise elastic tissues) and signaling for new blood vessels. Thus, the process of niche construction can involve a twisted form of cellular cooperation between cancer cells and apparently normal cells, in which cancer cells exploit the "willingness" of normal cells to "help." This is one of the most bizarre and fascinating occurrences within the tumor microenvironment—cancer cells hijack the normal cells of the body to promote the cancer cells' survival and proliferation.

One of the most important aspects of niche construction in cancer is the building of blood vessels that feed tumors. As cancer cells use up the resources in their local environments, their growth can be limited because of a lack of raw materials for building new cells.

The bloodstream is an excellent source of raw materials for building cells, and cancer cells evolve to recruit blood vessels to help fuel their growth when resources start to get scarce.

Scarcity drives competition, as well as cooperation. It drives exploitation, but also innovation. Humans, for example, have built sophisticated infrastructure systems to extract and deliver resources so that we don't have to deal with scarcity of resources on a day-to-day basis.

One of the most amazing examples of infrastructure is the irrigation system that the Hohokam started building in approximately 600 CE. The Hohokam are a group of Native Americans who lived along the Salt River in Arizona—the very place where I now live and work. Over the course of about eight centuries, they built an irrigation system with hundreds of miles of canals. Using digging sticks, they dug trenches, some up to twelve feet deep, to deliver water to households and farms where people otherwise wouldn't have been able to survive so far from the river. Just how they orchestrated the construction of this marvel of engineering—and how they managed this resource delivery system once it was in place—remains largely a mystery to modern archeologists due to the lack of written records.

The effective management of the canal system is even more of a mystery than its engineering and construction. Irrigation canals, like those that the Hohokam built, are brimming with social dilemmas, and social dilemmas can make it extremely hard for the actors in a system to effectively coordinate and cooperate. Everyone in the system may be tempted to take more than their fair share of water. Anybody upstream in the irrigation system can open up the taps and monopolize the water, leaving little for those downstream. In addition, the temptation exists to simply free ride on the efforts of those who work to make and maintain the system. This is a slightly different problem than cheating by just opening up the taps. Building and maintaining irrigation systems requires time and energy that some people have to put in, while anybody who is part of the irrigation system can benefit from being a part of it. This means that irrigation systems are a double whammy of social dilemmas—there is both a temptation to open up the taps as well as a temptation to free-ride on the effort of others to maintain the infrastructure. And so it is quite astounding that the Hohokam could manage and maintain this canal system for centuries.

Just like we have built irrigation canals to carry water to households that need it, during our development our cells essentially built an irrigation system to carry blood around the body. While we were in the womb, special cells called endothelial cells (the cells that form the walls of our blood vessels) invaded all the tissues of our bodies, creating a network of blood vessels that transports and distributes resources. But these blood vessels are not static—they constantly grow and change based on the signals they get from the cells around them. For example, healing signals can increase the blood flow in these vessels and even lead to the growth of new blood vessels. This system allows for the dynamic management of resources in the body, getting blood to the cells that need it when they need it.

Our body is a special kind of ecosystem that has been set up by the multicellular body during development to deliver resources to all the cells that make us up. When the body is functioning properly, resources are carried by the bloodstream to all of our peripheral tissues, where they feed the cells so they can have the energy they need to do their jobs and make us viable multicellular organisms. Cells rely on this resource supply to survive and function properly. When everything is working as it should, cells get the resources they need and our bodies thrive. This is not unlike a canal system that transports water to households that need it. Our bodies are like a well-functioning canal system with trillions of cellular households, successfully solving the social dilemmas that usually make cooperation on this scale impossible. But this solution is precarious, and if cells inside the body start taking advantage of the system, the fabric of the multicellular society can start to unravel. Cancer cells can threaten the local ecosystem and multicellular infrastructure in which they grow. Cancer cells tap in, extracting resources from the bloodstream, and tap out, draining the local resources until the infrastructure around them collapses.

In both of these systems—canal systems that distribute water and the vascular system that distributes blood around our bodies—cheating can be a problem. All of the social dilemmas that apply to irrigation systems apply to cancer cells as well. Cancer cells "open up the taps" by sending out signals that increase the permeability of blood vessels so that more nutrients can flow to them, leaving less for the cells that are downstream. In addition, cancer cells sometimes work together to signal for and build new blood vessels—like a group of people might get together to dig a trench off of a canal system to tap into the infrastructure that others have built. This cooperation can allow them to exploit the body more effectively. But this cancer cell cooperation is often short-lived. Any cancer cells that free-ride will have an advantage over the more cooperative cancer cells that incur the costs of creating new blood vessels.

When cancer cells start siphoning off resources for themselves, this process not only takes resources away from normal cells, but it threatens the integrity of the resource delivery infrastructure—which can lead to the collapse of blood vessels. A canal system can be perfectly stable when it is empty; however, when a blood vessel is empty, pressure from the tissues around it will cause it to collapse. When cancer cells open up the taps by signaling for more nutrients from the blood, less fluid will be in the blood vessels to maintain blood pressure and keep the vessel from collapsing. If a vessel collapses, the cells that were depending on it either die or, if they can, they signal for more blood vessels, building a new resource delivery infrastructure, which will again be vulnerable to being exploited and collapsing. This is one of the reasons blood supply in tumors can be so erratic and variable. Vascularized tumors are essentially a cluster of social dilemmas playing out along every millimeter of blood vessel, with cooperation emerging and collapsing constantly, mirrored by the emergence and collapse of the blood vessels that the cancer cells create and exploit.

Escaping from the Eco-pocalypse

The behavior of cancer cells can lead to a cellular tragedy of the commons. Cancer cells that replicate and use resources quickly have a short-term advantage over cells that are more restrained. Sometimes, cells that exploit their local environments will die, either because they have no resources left or because the amount of waste products around them overwhelms their ability to effectively detoxify. Other times, cells find a way out of this local ecological crisis, either by signaling for more resources, as we saw earlier, or by evolving to move and colonize new niches.

The core idea of dispersal theory in ecology is this: when organisms exploit their local environments, it can lead to selection for dispersal—for individuals that have the ability to move and find new environments to colonize. The very same principle applies in cancer evolution. When cells exploit their local environments, this leads to selection for cells that can move, and also for those that move more readily in response to poor ecological conditions.

My colleagues and I were inspired by this ecological principle and created a computer model of cancer cells to explore how the overuse of a resource might affect the evolution of cell mobility. Our computer model represented cancer cells living inside a simulated tissue with blood vessels delivering resources to them. In this model, there were cells that consumed resources quickly (like cancer cells do), and cells that consumed resources at a pace commensurate with the blood supply coming in (representing normal cells). We wanted to compare how cell mobility would evolve across these different conditions and see whether cells that consume resources quickly create ecological conditions that subsequently favor higher mobility and dispersal. As expected, we found that rapid resource use and environmental destruction by neoplastic cells drove the evolution of cell mobility. In accordance with dispersal theory, the high resource consumption of cancer cells leads to selection for cells that move.

Ecological dispersal theory has important implications for our understanding of invasion and metastasis. After invasion, and especially after metastasis, cancer is much harder to treat. Our model suggests that cells may evolve to move much earlier in progression than previously thought—long before invasion and metastasis are apparent. The model shows that cancer cells could evolve to move simply because they consume resources quickly and destroy their immediate environments. This is consistent with some curious facts about metastasis, like the fact that metastases often arise from cancer cells that left a tumor early in progression, in a process called "early dissemination." This suggests that evolution of cell mobility may happen early in progression, even though the consequences of this cell mobility may not be apparent until after the tumor has invaded and metastasized.

Dispersal evolution and niche construction may seem like diametric opposites: dispersal evolution happens as a result of environmental destruction, niche construction is a process of environmental creation. But they actually go hand in hand: environmental destruction leads to selection for dispersal, but effective invasion and colonization of new environments requires active niche construction and cooperation among cancer cells. In order for cancer cells to invade and get past the basement membrane, they must cooperate to construct a niche at the invading front that allows them to get through the body's barriers. As we saw earlier, the act of invasion often requires cancer cells to cooperate with each other to produce matrix metalloproteinases that break up the basement membrane. In addition, cancer cells can "trick" the vascular system into letting them through the endothelium (the membrane around the blood vessels) by coordinating their electrical signaling. Once they are able to get in and out of the bloodstream, they can then hitch a ride throughout the body and colonize new tissues and organ systems.

A Revolution of Cooperation

Cancer cells can exploit the ecosystem of the multicellular body and sometimes cooperate and coordinate their behavior to do so. We have seen how cancer cells can use cooperation to construct a niche that protects and feeds them. They also use cooperation to more effectively invade and metastasize, allowing them to colonize new tissues. How does cooperation among cancer cells evolve? What, if anything, keeps that cooperation stable in the face of cheating?

Cancer biologists are in the process of answering these questions, and we can use cooperation theory as a guide to understand cancer cell cooperation. Cooperation evolves based on the same fundamental principles across all systems, which means that we can apply the theoretical frameworks and models of the evolution of cooperation to cancer. This will be our starting point for discussing several possibilities for how cancer cells can evolve to cooperate with one another.

Let's go over a few possibilities for how cooperation could evolve among cancer cells. In evolutionary biology, when we see a particular trait (in this case, cooperation) it is common to ask whether it is an adaptation (something that resulted from natural selection favoring that trait), a by-product (a side effect of an adaptation), or just noise (something that didn't result from natural selection at all). In cancer, all of these explanations may hold. It is likely that some of the cooperation that we observe among cancer cells is just an evolutionary accident. Other examples of cooperation might be by-products of individual cell abilities like colonization adaptations. And other cases might be the result of natural selection favoring cooperation, a possibility that I delve into below. Later in this chapter I will come back to the question of whether noise and by-products can explain cancer cell cooperation. But now, let's take a deeper look at the question of when natural selection could favor the evolution of cooperation among cancer cells. What mechanisms might enable selection to favor cancer cell cooperation? Genetic relatedness among cancer clones? Reciprocal altruism and repeated interactions among cancer cells? How about fitness interdependence and multilevel selection inside the body?

Let's consider whether repeated interactions among cancer cells could contribute to the evolution of cooperation among them. As we saw in chapter 3, repeated interactions, such as reciprocal interactions in which individuals exchange benefits, is the most widely accepted explanation for the evolution of cooperation among nonrelatives. Repeated interactions can allow the benefits of cooperation to come back to the cooperator, making the strategy of cooperation more viable than that of cheating. Might different cancer cell clones use a reciprocal strategy to get benefits from interacting with one another?

Cancer cells do have interactions with one another that look very much like reciprocity (or at least mutualism). For example, sometimes one type of cell that produces growth factors benefits another type of cell that produces factors that help protect from the immune system. By cooperating, they can get the benefits from essentially dividing the labor of producing all these factors. Whether we should consider this to be by-product mutualism or consider it to be an example of "cellular reciprocity" is an open question. Typically, we think of reciprocity as a conditional strategy. In cancer, we don't know much about how cells can respond to each other. But it is certainly possible that cancer cells could evolve to cooperate conditionally—for example, only producing a public good if nearby cells are producing public goods as well.

Even if the jury is out on the question of whether cancer cooperation qualifies as reciprocity, it is clear that repeated interactions among cancer cells can set up conditions that can favor cooperation among cancer cells. And regardless of whether this is by-product mutualism or conditional reciprocation it is clearly an example of positive assortment. Positive assortment is when cooperators are more likely to interact with each other than with a random individual in the population—and positive assortment can favor the evolution of cooperation, regardless of whether the cooperation is among relatives, repeated interaction partners, or even members of different species who provide benefits for each other. So repeated interactions among cancer cells could shed light on the evolution of cancer cell cooperation in the context of division of labor.

Among evolutionary biologists, reciprocity is one traditional explanation for the evolution of cooperation; the other is kin selection via genetic relatedness, sometimes referred to as inclusive fitness (because it measures fitness *including* the fitness of kin, discounted by the degree of relatedness). Some cancer biologists are skeptical about the importance of kin selection as an explanation for cooperation among cancer cells, but it is a possibility that is worth seriously considering. As we have seen, cancer cells often exist in groups that are highly related because a mutation that confers an evolutionary benefit will often result in a clonal expansion—a group of cells that share that mutation. If cells in a clonal expansion share a mutation for something like production of growth factors, these factors can act as a public good that provides benefits for neighboring cells as well. If the neighboring cells belong to the same clonal expansion—and have the same gene for producing that growth factor—this genetic relatedness among clones provides an extra evolutionary boost for the gene that produces the growth factor. Even if the cell produces this growth factor at a cost, the genes coding for producing it could still be favored in the population of cells.

As we saw in chapter 2, natural selection favors genes that increase the survival and reproduction of other individuals who share copies of those genes. If a cell has a gene for producing a growth factor, and the cells that are nearby share the gene as well, that gene will benefit and expand in the population. This means that genetic relatedness among cancer cells may help explain some instances of apparent cooperation. But it is also possible for cooperation to evolve among cells that produce different growth factors produced by different genes, through the same process of positive assortment that I discussed earlier. It is not strictly necessary for the same genes to underlie cooperation for cooperation to be selected. Cooperation can evolve as long as cooperators preferentially interact with one another.

Cooperation among relatives has evolved countless times in nature. Sometimes this cooperation with kin evolves along with the ability to recognize kin. We can see this in organisms such as humans, who have very long periods of parental investment during which we take care of our offspring (and often other relatives as well). But kin recognition is not actually required for cooperation among kin to evolve. In organisms with short periods of parental investment, relatively low mobility, and low sociality, it isn't necessary to recognize who is and who is not kin for investment in relatives to be effective. For example, if offspring stick around, parents who simply provide benefits for anyone near them will end up investing in their offspring just because they are close. Cooperation among relatives can evolve via kin selection when there is a kin structure (like offspring staying near parents) in which recipients of benefits are likely to be relatives.

Kin structure is one potential way that cooperation among more related cancer cells could evolve, even if those cells have no ability to recognize one another. Cells growing in a cluster of related clones are simply more likely to interact with a highly related clone than a less related cell, creating conditions under which cooperative capacities could be selected through kin selection.

My colleagues and I have suggested that cooperation among genetically related cancer cells may play an important role in cancer evolution—and we used a model to help explain another puzzle in cancer biology: the existence of "cancer nonstem cells" (in other words, cancer cells that have limited reproductive potential). Tumors often have very high proportions of these nonstem cells, with limited potential to divide. Cancer nonstem cells cannot propagate the tumor because they reach a replicative dead end. The limited potential of these cells is a puzzle from an evolutionary perspective. Cancer cells should evolve to have *unlimited* potential for division because—all else being equal—cells that can make the most copies of themselves will be the ones that leave the most cellular descendants. But this is not what we see when we look at tumors—experiments suggest that somewhere between 75 percent and 99.999 percent of cells in tumors are cancer nonstem cells and cannot propagate the tumor. How are these cells maintained in the population of cancer cells if they don't leave cellular descendants?

We created our model, including cancer nonstem cells with limited potential to divide that are able to provide a fitness boost to genetically related cells. When we included such cells in our model, we found that these cells with limited potential to divide were maintained in the population of cells. If this process is indeed occurring in tumors, it suggests that the dynamics might be similar to what happens in some species that engage in what is called collective breeding, in which some individuals reproduce while others act as helpers (this phenomenon occurs with many birds who have related helpers at the nest). This kind of system is possible because of high relatedness among the individuals in the breeding group.

Another potential parallel with groups of cancer cells may come from social insect societies—in these societies some individuals are able to reproduce while others are not. Many social insects have sterile workers who do not reproduce and queens that do. It is possible that cancer cell groups might recapitulate some of these differences in reproductive capacity among members. Interestingly, social insect societies are often able to thrive in harsh environments. Being social makes sense when resources are unpredictable, because sociality provides a buffer against the challenges of the variable environment. Like groups of cancer cells, social insect societies are very successful at invading new territories and colonizing new environments. In fact, some social insects, including certain species of ants, are so good at invading and colonizing that they have become problematic pests and can even threaten conservation efforts because they deeply change the structure of the ecological community around them, often displacing native species. This may be a parallel with cancer cells that change the structure of the ecosystem in the body and displace the "native species" of normal cells. Cancer cells evolve to be highly successful at invading and colonizing new environments in the body during late stages of cancer. It is very much an open question whether colonies of cancer cells can evolve to have a similar structure and functionality as social insect colonies, but the connections are intriguing and deserve further study.

Environmental conditions greatly affect cooperation across all systems—from those of humans to honeybees to cancer cells. Cooperation makes sense in harsh and challenging environments where survival is impossible without it. During cancer progression, cancer cells face many environments in the body where they can't survive without cooperating. For example, when the blood supply has been exhausted locally, cells may be doomed unless they can cooperatively signal for blood vessels. Cooperation can also help cancer cells invade neighboring tissues and metastasize.

One of the best pieces of evidence that cooperation is important in cancer progression comes from the finding that cancer cells often metastasize as a group, and that larger clusters are more successful at metastasizing. These cancer cell clusters circulate in the blood during late-stage cancers and they can be measured through taking blood samples (figure 6.1). The size of these clusters makes

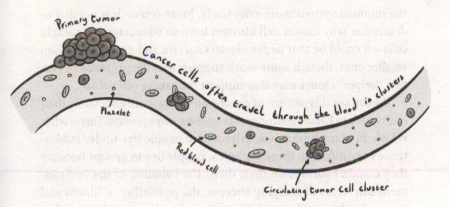

FIGURE 6.1 Cancer cells can travel through the blood as single cells or as clusters of cells. Studies show that cancer cell clusters can metastasize more effectively than single cells circulating in the blood.

a difference for patient survival. Breast cancer and prostate cancer patients with circulating tumor cell clusters do not survive as long as patients who have only single tumor cells circulating in their bloodstream. Clusters of cancer cells seem to be better able to colonize the body than individual cancer cells. In a mouse model of mammary tumors, circulating tumor cell clusters were twenty-three to fifty times more likely to successfully create metastases than individual cells. There is also evidence that certain polyclonal tumors (containing multiple different clones) can have a proliferation advantage, because they contain clones that can collectively colonize and maintain the cancer niche.

Together these findings suggest that metastasis—and the colonization of new environments that is required for metastasis—may be much easier for colonies of cancer cells than for individual cells acting alone. It might be that in some cases, colonies of cancer cells actively cooperate to colonize new environments while other cancer cells may simply passively benefit from being in a group. It is possible that these clusters have an advantage in niche construction once they reach a new environment, producing growth factors for each other, signaling for blood vessels, and perhaps even hiding from the immune system more effectively. More research is needed to determine why cancer cell clusters have an advantage over single cells—it could be that larger clusters survive and grow better than smaller ones, though some work suggests that the presence of certain "helper" clones may also support the cancer cell colony.

We can see this pattern of groups surviving more effectively than individuals in harsh environments across many systems. This is what we see in many human societies where people live under subsistence conditions. In these societies, people live in groups because they couldn't survive on their own. The volatility of the environment—in terms of foraging success, the possibility of illness and injury, the chance of natural disasters, and extreme weather—makes it hard to survive as an individual human. Similarly, social insects like honeybees and ants have evolved to live in large colonies that help buffer them from the harsh environments they live in, in addition to enabling division of labor on a massive scale. The similarities among cancer cell colonies, human groups, and honeybee colonies are striking and suggest that we can gain further insights into the evolutionary dynamics underlying cancer colonies by considering the pressures that shaped colony living in other species.

Generally speaking, living in a large group helps individuals pool risk, which can increase their chances of survival in challenging conditions. In these situations, individuals become more dependent on one another for survival and reproduction, leading to greater fitness interdependence. Fitness interdependence is when individuals are dependent on one another to get their genes into the next generation. Fitness interdependence in humans often happens in situations where the survival or reproductive success of individuals is yoked together (such as mating relationships in which the survival and success of mutual offspring is at play), in periods of war (when soldiers are dependent on one another for survival), and in harsh and unpredictable environments (where it may be impossible to survive as a loner). Among cancer cells, fitness interdependence likely happens when clusters of cells land in new and harsh environments where survival would be impossible without cooperation.

If the only way to survive is by means of mutual cooperation, then cheaters will end up destroying themselves unless they can quickly find a cooperator to exploit. It is possible that the benefits of cooperation in harsh environments select for cancer cell cooperation at the frontiers of tumors, especially during invasion and metastasis. In the next section, we'll look at this very question of whether cooperation among cancer cells can be favored by natural selection during metastasis.

Metapopulations and Metastasis

Social insect colonies evolved to have high levels of cooperation—including sterile worker castes—for several reasons. They are often comprised of individuals who are highly related to each other, making it possible for any genes coding for workers' dedication to the queen's reproduction to get passed along. (Many social insects are haplodiploid: males have no fathers and subsequently have half of the genetic material that females have. This means that, if all workers share the same queen and that queen has only mated once, workers will be related by three-quarters, rather than the typical one-half for siblings.) In addition to high relatedness making cooperation more viable, the population structure of social insect colonies also allows for more cooperative colonies to outperform less cooperative ones. These colonies can form a population of groups—a metapopulation—and so cooperative behavior can increase in the population as a whole even if, within each group, cheaters get a higher payoff. This is the process of multilevel selection that we discussed in chapter 3, wherein selection can operate on different levels simultaneously (for example, selection can act on the level of the individual favoring cheating while also acting on the level of the colony to favor cooperation).

Some cooperation theorists categorically assert that nothing can evolve "for the good of the group," including cancer cells in a cancer cell cluster, because evolution will always favor cheaters within the group. But the fact is that cooperation with groups can sometimes evolve despite the evolutionary advantage of cheaters within the group. Groups differ in how successful they are: some die out quickly while others thrive and perhaps even send off "buds" of new groups (or what are called metastatic cascades in the case of cancer). Since groups with cooperators fare better than groups with cheaters, cooperation can increase overall, at least temporarily.

In order to see whether the process of metastasis can select for cooperation within cancer cell colonies, first we must look at the traditional models for the evolution of metastasis and examine their limitations. Then we will turn back to the question of whether multilevel selection might be operating during metastasis to favor cooperation within cancer cell colonies.

There are two accepted models of what happens over time in metastasis (figure 6.2). Unfortunately, neither of them matches current data about what metastasis actually looks like. One of the accepted models is the linear model, which assumes that metastasis happens late in the game of tumor evolution, and that metastases are always driven by the most "advanced" clone. In the linear model, each metastasis can be placed in a linear order with the first metastasis coming from the primary tumor, the second metastasis coming from the first metastasis, the third metastasis coming from the second metastasis, and so on. This model is an extension of early models of somatic evolution, which assumed that cancer proceeds in stages as each mutation builds on the last one in a step-by-step fashion.

The other currently accepted model is the parallel model. You could say this model is more ecological compared to the linear model, in that it envisions an environment in which the "seeds" of metastasis are planted early on during tumor dissemination. This model assumes that all metastases originate from the primary tumor, with some originating early in progression when no metastases may be detectable and some later on. The saguaro cactus sends out hundreds of seeds to propagate; the parallel model posits that a primary tumor does the same, sending out seeds for metastases early in the process of cancer evolution. Some of these seeds land in places with rich resources or are lucky enough to mutate in ways that allow

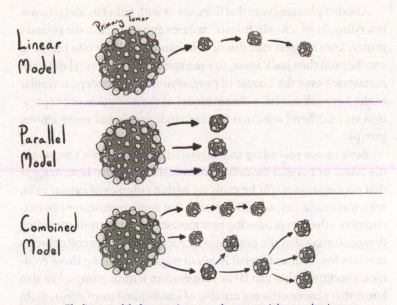

FIGURE 6.2 The linear model of metastasis assumes that metastasis happens late in tumor evolution, and that the first metastasis (and any subsequent metastases) are driven by the "most advanced" clone. The parallel model of metastasis assumes that the "seeds" of metastasis are planted early on during a process called tumor dissemination, and that metastases all originate from the primary tumor. The reality is likely to be a combination of these two models, in which there is a combination of linear and parallel processes. Adapted from Turajlic and Swanton 2016.

them to better signal for blood vessels or evade the immune system, and so they flourish.

Despite the appealing simplicity of these two models, neither of them fits with the current data. There are elements of both models in the data—even within one sample of one cancer. For example, metastatic cascades involve elements of both the linear and parallel models for metastasis. In metastatic cascades, a primary tumor gives rise to many metastases, but only some of these metastases can then generate further metastases. In other words, some metastases are more evolutionarily successful (at the level of metastatic cell colony) than others, budding off new propagules that can go off and successfully colonize new areas of the body.

Another phenomenon that does not fit with either model is tumor reseeding, in which cells from metastases appear back in the primary tumor. Data suggest that tumor cells can travel from one tumor to another and then back again, or even spend time in several different metastases over the course of progression. This concept is similar to the haystack model, a classic model of the evolution of cooperation via multilevel selection in which individuals can move among groups.

Both tumor reseeding and metastatic cascades don't fit within the linear or parallel models of metastasis. They also both suggest that cooperation might be evolving within colonies of cancer cells, with metastatic cascades illustrating that some metastases can outcompete others in producing new metastases, and tumor reseeding demonstrating that the population structure of cancer cell colonies involves low (but non-zero) levels of migration among these colonies, conditions that can favor cooperation within groups. We also know that cancer cells are capable of conditional movement, making it even more likely that selection will favor cooperation within these cellular groups.

If cancer cells are evolving to cooperate within tumors, this would dramatically affect our understanding of metastasis. For example, entire generations of invisible micrometastatic colonies, undetectable by our current technology, could be evolutionarily selected for colony-level cooperative properties. In this scenario, colonies that could best cooperate to grow new blood vessels, evade the immune system, and of course "reproduce," budding off new metastatic propagules, would be the ones most likely to contribute to the next generation of cancer cell colonies. We don't yet have direct evidence for this scenario, but we can look more closely at whether it is possible for colony-level cooperative traits to evolve. In order to do that, we must return to the question of whether multilevel selection could be operating on cancer cell colonies during metastasis.

Recall that multilevel selection is simply a case where natural selection is acting on more than one level at the same time (e.g., at the individual level and at the group level). If we want to know whether colony-level cooperation could evolve, we need to examine whether the conditions for natural selection are met at the level of the cancer cell colony. These conditions for natural selection are variation among the colonies, differential fitness among colonies (i.e., different rates of survival and/or budding off of new cellular groups), and heritable differences among the colonies.

Let's look at the first criterion for natural selection among colonies: variation. Do cancer cell colonies vary? Indeed they do. During metastasis, colonies of cancer cells vary genetically. How about the second condition: Is there evidence for differential fitness among cancer cell colonies? Yes. According to current phylogenetic trees of cancer metastasis, the cancer cell colonies have different rates of survival and creation of new colonies. This is part of the idea of metastatic cascades that I talked about earlier, where some cell colonies give rise to many new colonies, whereas other colonies do not appear to spawn new ones. Not only is there evidence for differential fitness, there is also evidence for direct competition among colonies of cancer cells inside a host: for example, sometimes when a large primary tumor is removed, tiny metastases will grow rapidly because the primary tumor is no longer monopolizing nutrients and producing inhibitory factors. This phenomenon of primary tumors suppressing metastases, known as "concomitant tumor resistance," has been widely observed in both animal experiments and in human patients. To summarize, it is clear that cancer cell colonies meet the first two criteria for natural selection to be operating at the level of cellular groups: there is variation among those groups and differential fitness, with some groups "reproducing" more than others.

What about the third condition for natural selection, heritability? Do cancer cell colonies inherit traits from their "parent" colonies? This is where open questions still remain. Current methods don't allow scientists to track multiple generations of metastases with enough detail to determine whether daughter colonies look like their parent colonies in terms of growth rates, survival, or other relevant traits. If we find evidence for heritability in cancer cell colonies, this would show that groups of cancer cells can act as units of selection and that the cells within these colonies are under evolutionary pressure to cooperate in ways that could make those cancer colonies more effective at metastasizing.

There are already several pieces of data that suggest that cooperation within cancer cell colonies could be an important driver of metastasis, but there are also many open questions. We know that clusters of cancer cells often colonize together—and with greater success than individual cells. But we don't know why these clusters do better than individual cells. It might be that the ecological challenges of successfully migrating to and colonizing a new tissue selects for clusters that are better at metastasizing. It could also be that clusters of cancer cells evolve to have some reproductive division of labor, with some cells proliferating while other cells support the proliferating cells, almost like a protomulticellular organism. It is clear that cancer cells can promote one another's fitness in various ways, including producing growth factors and survival factors, and helping to hide one another from the immune system; but we don't know much about how these cooperative abilities among cancer cells may be contributing to metastasis.

These possibilities and open questions motivated me to team up with several colleagues to write a paper entitled "The Darwinian Dynamics of Motility and Metastasis," in which we take the principles of evolution, ecology, and cooperation theory and apply them to metastasis to develop several hypotheses about what might be happening during progression to metastasis. Based on current data, we concluded that the process of metastasis could select for colonies of cancer cells that have some of the same traits that organisms have: life cycles with distinct phases of growth and reproduction, as well as life history strategies. Some cell colonies may have fast life history strategies, generating lots of new cell colonies early, and others may have slow life history strategies, putting out fewer but more viable propagules.

My colleagues and I are not the only people who have suggested that multilevel selection could operate on metastases to favor cooperation within them. The philosopher of biology Anya Plutynski notes that metastasis is a complex process that requires cooperation among tumor cells. She rightly points out that most incipient metastases fail before they even have a chance to take hold, getting destroyed in the circulatory system or simply failing to colonize. She suggests that "some metastatic populations are more effective at colonization than others," and that some metastases will be "more or less successful at disseminating secondary metastases." In other words, metastatic cancer colonies that are best at colonizing and disseminating secondary metastases would have an advantage over other cancer cell colonies (and individual cells, for that matter), and this would lead to selection favoring colonies that can quickly and effectively colonize new environments and send out new metastases.

But not all philosophers of science agree. Others have argued that multilevel selection is not relevant to late-stage cancer because metastases are unlikely to "reproduce" in a way that reliably leads to heritability of variation among them. As I mentioned earlier, this issue of heritability among cancer colonies is still an open question. It is the one condition for natural selection that we don't yet have evidence for among cancer cell colonies. Although to be fair there also is no evidence against heritability at the colony level. And if we look at other populations that form colonies, like social insects, there is evidence for heritability of traits at the colony level. This suggests that we should take seriously the possibility that heritability of traits can occur at the cancer cell colony level, and that researchers should seek to measure heritability of traits at this level.

There are still many open questions about exactly what happens during metastasis because current technologies do not allow us to see how the population structure of cancer cell colonies changes as tumors become metastatic. Metastases can be microscopic, and our methods only allow us to detect metastases of about one million cells. We don't yet know how many steps there might have been in any given metastatic cascade before we can detect a metastasis. There could be orders of magnitude more than we can detect with modern imaging techniques: tens, hundreds, or even thousands of generations of evolution could be occurring among tiny metastatic cancer cell colonies. It may be that colonies of cancer cells are selected over multiple generations to be effective at extracting resources from the body, and this could be happening before the metastases grow to a size that can even be detected by modern imaging technology. If there are indeed many generations of cell colonies competing with one another to effectively exploit the body, this opens up the possibility that sophisticated colony-level phenotypes could evolve over these many generations. For example, colonies might be under sustained selective pressure to evolve coordinated evasion of the immune system, effective angiogenesis to get resources, cooperative growth signaling, and the ability to send out new metastatic propagules that may continue the cascade.

It is also difficult to know how many tiny metastases could be competing with each other early in metastasis. Earlier I discussed how cancer cells that leave a tumor early in progression will sometimes give rise to metastases much later, a process called early dissemination. We also know that large tumors can effectively suppress the growth of smaller tumors. But it remains a mystery how big tumors must be in order to dominate nutrients and produce inhibitory factors that allow them to effectively dominate other tumors in the host's body. Future work in animal experiments (for example, with mice) could answer some of these questions. Gene expression data can help us learn more about factors that cancer cell colonies are producing that allow them to compete with each other. Gene expression data can also provide some hints as to the kinds of collective phenotypes that can emerge in cancer. But gene expression data can't reveal what kinds of collective phenotypes a colony of cancer cells would have in the ecological context of the body because of all the complex interactions that happen with other cells and the tumor microenvironment. And both of these factors—the number of cell colonies and their collective phenotypes—will affect the strength of selection on cancer cell colonies to be more effective at metastasizing.

As we learn more about the genetics and evolution of metastasis, a fascinating and possibly alarming picture is emerging: colonies of metastatic cells may be evolving in their own right. There are hints that, during metastasis, a metapopulation structure emerges that could select for cell colonies that are good at metastasizing. These cell colonies may give rise to metastatic cascades of cell colonies that are most successful at exploiting the host. Multilevel selection could be operating on cell colonies as units of selection, which could potentially favor greater and greater colony-level metastatic abilities as cancer progresses.

If multilevel selection favoring cooperative cell colonies is ongoing during cancer progression, it might explain why cancer is increasingly hard to treat as it progresses. If cancer cells can capitalize on the benefits of cellular cooperation and evolve colony-level phenotypes that help them survive and thrive throughout the body, they might be extremely difficult to eliminate. If metastasis is a result of cooperative cancer cell colonies, this could also help to explain why no gene has been found for metastasis, nor has a particular gene pathway implicated in it. The idea that metastasis is driven by cooperation within cancer cell colonies remains speculative, but as technology advances so will our ability to test whether cancer cell colonies evolve to utilize multicellular cooperation and what the implications are for how we should treat cancer.

We need to better understand metastasis itself in order to treat metastatic cancer effectively. How we approach treatment depends on our understanding of cancer, and it is clear that there are still several critical holes in our knowledge. We have seen how multilevel selection could be at work during cancer progression (if there is high enough heritability at the colony level). One important implication of this is that, if selection among metastatic colonies is the mechanism that pushes forward cancer, then removing the primary tumor would not stop a metastatic cascade (and, in fact, there is evidence that removing a primary tumor can sometimes harm the patient). If metastasis is driven by cancer cell cooperation, it may be wise to focus our efforts on disrupting cooperation/coordination in the metastases so that they don't keep growing or producing more propagules. I will return to this and related ideas in the final chapter of this book.

By-products, Accidents, and Other Explanations for Cooperation

So far, we have focused on the possibility that cooperation among cancer cells in cancer cell colonies could arise as a result of natural selection favoring cell cooperation. But we also must consider two other possibilities: (1) that cancer cell cooperation could arise as a result of by-products of other things the cell is doing and (2) that cooperation among cancer cells might simply be an accident.

Cooperation is considered a by-product if it happens as a result of adaptations that evolved for other reasons. In the case of cancer cells, cooperation could be a by-product of other things that cancer cells have been selected to do at the cell level, like move, colonize, or evade the immune system. Some of the things cells need to do to be good at surviving and colonizing may inherently provide benefits to other nearby cells. For example, any kind of niche construction/improvement of the environment could create a public good that could be utilized by nearby cells. If a cell produces factors that break down the extracellular matrix, allowing them to invade more effectively, the cells behind them can benefit from the cleared path (like humans following a trail through the forest, cut by those who traversed it before them). In addition, cells that signal for blood vessels can bring resources to a region where all nearby cells can benefit. It may be that some examples of apparent cooperation between cancer cells are a result of by-products of cells doing what is in their cellular best interests.

Sometimes by-product benefits can flow in two ways simultaneously, giving rise to what looks like a coordinated cooperation—but it may be coincidental. For example, if two populations of cancer cells with different mutations (say, one that produces a growth factor and another that produces a factor that allows for invasion) happen to be near each other, they may end up providing benefits for one other, without any active coordination or conditional "reciprocity." This situation is called by-product mutualism. In harsh environments where it is difficult to survive without producing these kinds of factors, by-product mutualism becomes even more likely because "cheating" is not really a viable option. These kinds of by-product benefits are likely a part of the explanation for why we see cancer cells cooperating inside tumors. However, this by-product benefits explanation can turn into a natural selection explanation—if, for example, the cells that provide these benefits for one another end up interacting preferentially as a result of

spatial proximity or other factors that promote positive assortment, and these cooperators get an evolutionary advantage over
noncooperators.

Within any cooperative group, cheaters always have an advantage, so cancer cell cooperation—when it arises—will be vulnerable
to cheating as well. This is the reason we are susceptible to cancer in
the first place: the cells in our body evolve to cheat and take advantage of multicellular cooperation. But the advantage that cheating
cells have over cooperative cells can potentially undermine cooperation *among* cancer cells as well.

This suggests that clusters of cooperative cancer cells could be
fragile and short-lived unless there are some mechanisms for enforc-
ing that cooperation among the cancer cells. Without some enforcement mechanisms, cooperation will exist in only a transitory state.
This leads to a reasonable null hypothesis for the evolution of cooperation among cancer cells: Cooperation among cancer cells might
arise as a result of a random process, like genetic drift. After this,
cooperation may be at a disadvantage in the cancer cell population.
This possibility—that any cancer cell cooperation arises accidentally
and will be short-lived—is a reasonable starting point for an explanation for the evolution of cooperation among cancer cells, unless we
have other evidence.

But even if cooperation among cancer cells is just accidental and
transitory, it is still possible for cancer cell cooperation to have very
important effects on progression. For example, we have seen how
cooperation among cancer cells is important for successful invasion
and metastasis, processes that open up new environments for cancer
cells to colonize. If cooperation among cancer cells helps facilitate
invasion and metastasis, this could have major impacts on cancer
progression—even if that cooperation is short-lived and breaks down
after the cancer cells have invaded or colonized the new environment.

Cancer cell cooperation might often be ephemeral, but if it happens at critical junctures, it could allow the cancer cells to invade
new tissues and metastasize throughout the body. We know that
cancer cells can use cooperation and coordination to invade tissues;
for example, they can use electrical and chemical signaling to move

together as a small group or even as a long conga line of cells through tissues and membranes. In addition, when clusters of cancer cells colonize in groups, they are more successful than when they are alone. With this information in mind, even short-lived cooperative groups of cancer cells could spur tumor progression, having a major impact on the ecosystem of the body and the health of the multicellular organism harboring them.

That said, there are reasons to think that cancer cell cooperation may not just be accidental and ephemeral. We have seen how some of the mechanisms that select for cooperation and help maintain it (like genetic relatedness and repeated interactions) can work among cancer cells. It is also possible that multilevel selection is acting among metastases, selecting for cancer cell colonies that are more effective at growing, surviving, and competing with other metastases.

Microbial Mediators

So far we have seen how cooperation among cancer cells can give them an evolutionary advantage within the ecosystem of the body. But there is another important piece of the somatic ecosystem that we have not yet looked at: the microbiome. The microbiome refers to all the microbes—including bacteria, yeast, and viruses—that live inside us and on us. Microbes have been found in and around tumors, and there is evidence that some microbes can contribute to cancer, and other microbes can help protect against it.

Broadly speaking, we can think of the microbiome as another potential player in the somatic drama of multicellular cooperation and cheating. It can either contribute to the multicellular cooperation of the body, or contribute to the breakdown of that cooperation. Some microbes are beneficial to humans and contribute to our health and well-being. They protect us from disease, help us process nutrients, and even lower the risk of inflammatory diseases and depression. We can think of these microbes as cooperating with us, their multicellular hosts, in ways that are beneficial for both us and them. But microbes such as those associated with infectious diseases do not usually have interests that align with our own. They thrive by exploiting us—using our resources for their own survival and proliferation.

Pathogenic microbes can make us sick in many different ways. One of those ways is by increasing our risk of cancer.

About 10 to 20 percent of human cancers are associated with specific microbial species, and many other microbes (as well as multicellular parasites) are suspected to play a role in cancer risk, even if it is indirect. This is not just the case for humans—many cancers in wildlife are associated with microbial infections. Sometimes microbes can directly benefit from cancerous cell proliferation, if the cancer cells are providing some benefit to the microbes. Ideally, our microbes are cooperating with our normal cells to help keep us healthy and cancer-free. But what happens if microbes cooperate with cancer cells instead of with the normal cells in our bodies?

Cancer cells and microbes can cooperate with one another, teaming up in order to better exploit the multicellular body. Microbes can provide benefits to cancer cells, and cancer cells can provide benefits in turn for microbes. This multispecies cooperation can evolve simply because of assortment—the preferential interaction of cooperators with one another—which I talked about earlier in this chapter. Cancer might not just be a problem of cancer cells cheating normal cells or even a problem of cancer cells cooperating to better exploit the multicellular body. Cancer might also be a result of microbes cooperating with cancer cells to help them thrive.

How exactly do microbes cooperate with cancer cells in ways that influence cancer risk? Some microbes, like the human papillomavirus virus (HPV), are quite direct about it: they get into the nucleus of the cell and increase cell proliferation, which increases the risk of cancer, in part by interfering with the p53 protein. This excess proliferation benefits the virus while it increases the fitness of the cell carrying it; both virus and cell make more copies of themselves. There are more subtle ways that microbes can increase cancer risk, including producing genotoxins that damage DNA and producing virulence factors that increase cell proliferation. Microbes and cancer cells can also produce growth factors for one another and they have the ability to protect one another from the immune system. The list goes on:

microbes can also help cancer cells to invade and metastasize by producing toxins that transform cancer cells from more sedentary to more motile, and producing quorum sensing molecules that contribute to metastasis.

But some microbes protect us from cancer, and some have even been used as part of cancer treatment. Microbes and microbial products have been used in cancer treatment for more than one hundred years and are still in use today. Doctors use the bacterium *Mycobacterium bovis* BCG in the treatment of bladder cancer, for example. There are many different ways that microbes and their products can help treat cancer, including activating the immune system, inducing cell death, and inhibiting new blood vessel growth. Microbes can also influence the success of cancer treatments—experiments have found that mice with their commensal microbiomes intact have better responses to therapy than mice that have been given antibiotics.

Microbes can enhance intestinal barrier function, improve immune function, inhibit cell proliferation, and help regulate metabolism (figure 6.3). Some studies suggest that probiotics and prebiotics may have cancer preventative effects in humans—a meta-analysis found that consumption of a lot of fiber (which is a prebiotic, since it feeds beneficial microbes) was associated with lower colon cancer risk—research in this area is new and not all studies find a protective effect, but the finding is intriguing. This is an active area for research and quite an exciting one. If we can more effectively prevent cancer and treat it using prebiotics and probiotics, we may be able to improve human health while reducing the cost and toxicity of our cancer treatments.

Microbes can also conditionally respond to information while they are inside our bodies. For example, they can turn on virulence genes in response to low nutrients. More work needs to be done to understand what role information processing plays in creating and stabilizing cooperation between the cells of the human body and the cells in our microbiome—in other words, whether they are engaged in a sort of reciprocal arrangement with one another. But it is clear that cooperation between our normal cells and our microbial cells is essential to our health—and that in some cases cooperation occurs between cancer cells and microbes during progression.

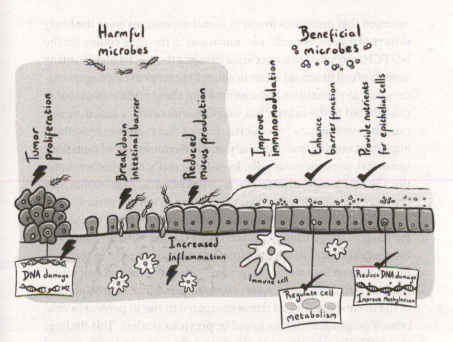

FIGURE 6.3 Microbes can influence many aspects of cancer progression, sometimes promoting cancer (*left*) and other times protecting against it (*right*). Harmful microbes can cause DNA damage and base pair shifts, increase cell proliferation in the tumor, increase inflammation, and also interfere with proper barrier function in the gut. Beneficial microbes, on the other hand, can enhance barrier function in the gut, improve immunomodulation, and help regulate cell metabolism. Beneficial microbes can also provide important nutrients and factors for epithelial cells that allow them to function better, reducing DNA damage and helping to keep methylation intact.

Some Clonal Expansions Can Halt Cancer

Throughout this book, I've focused on how cellular cheating can give cancer cells an evolutionary advantage over normal cells and lead to clonal expansions. But new work suggests that there might be an additional reason why some clonal expansions might happen: they might sometimes actually protect us from cancer.

We have seen how, over the course of our lifetimes, we accumulate mutations, some of which may increase our risk of cancer. In fact, we all have patches of skin with clonal expansions of cells that have cancer mutations—like mutations in *TP53*. Until recently, researchers assumed that mutations found in clonal expansions were the likely drivers of cancer. For example, mutations in the gene coding for the NOTCH1 receptor (an intercellular signaling protein involved in many aspects of cell function) occur in about 10 percent of esophageal cancers, and so researchers had assumed that these mutations probably contributed to the cancer. But Inigo Martincorena, a bioinformatician and evolutionary genomicist, realized that there was something important missing from this analysis: the measurement of mutations in normal, noncancerous cells. In order to really see whether mutations like these were driving cancer, it was essential to show that they are more common in cancerous tissues than in normal tissues.

So Martincorena and his team sequenced 844 small samples of esophageal tissue from deceased organ donors with no history of esophageal cancer to see how frequent mutations were in this normal, noncancerous tissue. Surprisingly, they discovered that mutations in *NOTCH1* were far *more common*—present in 30 to 80 percent of the normal esophageal tissue compared to the 10 percent in cancerous esophageal tissues found in previous studies. This finding, that *NOTCH1* mutations are more strongly associated with normal esophageal tissue than with esophageal cancer, has been replicated. These results suggest that clonal expansions of *NOTCH1* mutations might be protective against esophageal cancer. Indeed, in this same study, Martincorena and his colleagues found that *TP53* mutations were not common in normal esophageal tissue, though they were very common in cancerous esophageal tissue (in about 90 percent). It is possible that clonal expansions of *NOTCH1* mutations literally took up space in the tissues, making it harder for clonal expansions of *TP53* mutations to grow.

These results are important for several reasons. First, they challenge us to think differently about clonal expansions and cancer. We can't just assume that clonal expansions are all bad, and we can't assume that mutations that are common in cancer are necessarily carcinogenic. Martincorena's work suggests that some clonal expansions might actually be good, and some mutations might confer protection from cancer.

This raises an intriguing possibility: our multicellular bodies could have evolved to "strategically" generate clonal expansions to prevent cancer cells from expanding. This idea has been suggested by cancer biologist James DeGregori and his colleague Kelly Higa. They propose that *NOTCH1* mutations might occupy "decoy fitness peaks"—local peaks in the fitness landscape that prevent the population of cells from evolving in a more damaging direction. They argue that this might be part of the "program" that multicellular bodies evolved to minimize cancer risk. Think of an army preemptively occupying territory so that nobody else can do so—this is what DeGregori and Higa are proposing might happen with *NOTCH1*. And it wouldn't be the only case of clonal expansions being leveraged to protect us from threats—as we saw in chapter 4, our immune system uses somatic evolution to create clonal expansions of immune cells that help us fight infection and cancer.

If clonal expansions can sometimes help protect us from cancer, it suggests a novel approach to prevention, risk stratification, and treatment. For example, it could be possible to generate noncancerous clonal expansions with the goal of cancer prevention or preventing recurrence post-treatment. It might also be possible to measure existing noncancerous clonal expansions to assess risk of progression and improve monitoring in precancerous conditions.

If our bodies evolved to generate preventative clonal expansions, what mechanisms might underlie this ability? One mechanism could be mutation "hotspots"—regions of the genome that tend to mutate first because they are vulnerable to DNA damage during times of cellular stress. Multicellular bodies might have evolved to have certain mutation hotspots that can generate clonal expansions that take up space so that more dangerous mutants can't take hold. Mutations can be induced by cell-level stresses like DNA damage; some clonal expansions could be the result of an adaptive cancer suppression system that actually leverages this process to protect us from cancer.

That cells might occupy ecological space to crowd out more dangerous cells is analogous to our thinking about the role of beneficial microbes in helping to maintain our health: some microbes can be beneficial to humans simply because they can occupy the ecological space in and on our bodies so that disease-causing microbes can't.

Selfish Genes in Cancer Evolution

Other important players in the drama of cancer are the genes inside the cell—in particular, selfish genetic elements like transposons. Throughout this book, I have focused mostly on cells as the unit of selection for cancer evolution inside the body. Earlier in this chapter I discussed how cancer cell colonies can sometimes be units of selection. Now let's go small and look inside the cell at how genes themselves can be units of selection and how they could be playing an important role in cancer evolution.

We saw in chapter 4 how the differing genetic interests of our mothers and fathers can play out through epigenetics—with paternally expressed genes upregulating growth and maternally expressed genes constraining it. This conflict between maternal and paternal genes within the genome (known as intragenomic conflict) can influence our risk of cancer—but it is just one example of how conflict within the genome can manifest in cancer.

More generally, genes inside the genome can sometimes work at cross-purposes to one another, promoting their own replication at the expense of the cell or changing the expression state of the cell in ways that improve the gene's fitness. Just as organisms evolved to constrain cellular cheating during the transition to multicellularity, genomes evolved the ability to constrain gene-level cheating during the transition from a world of self-replicating DNA, in a free-for-all, to a world in which DNA replication is coordinated as part of a genome that is organized into chromosomes. This was one of the critical transitions during the evolution of life—it allowed for cooperation and coordination of genes within a genome, allowing cells to develop complicated behaviors that free-floating pieces of DNA never could.

But this genome level cooperation is by no means perfect. Even in a normal cell, pieces of DNA are jumping in and out of our genomes. Some pieces of our DNA are perfectly capable of copying themselves without waiting for the whole genome to get copied. And some of these bits of DNA—mobile genetic elements called transposons and retrotransposons—can copy themselves and move around the genome, inserting these copies back into the genome in new places. (Transposons copy themselves directly and insert back into the genome; retrotransposons are first transcribed into RNA and then back to DNA, which is then inserted in the genome.) According to evolutionary biologists Jonathan Featherston and Pierre Durand, mobile genetic elements "are functionally analogous to the presumed ancient replicators that cooperated to form primitive protein coding genomes" and, hence, their replication outside of the constraints of cell replication is essentially "cheating" in the cell-level DNA replication process. They are a reversion to a pregenomic lifestyle, just as we can think of cancer cells as a reversion to a unicellular lifestyle.

Transposons and retrotransposons make up a huge portion—almost half—of our genomes. Why are they there? The main reason is because they excel at replicating themselves. Our genomes have evolved to keep mobile elements under control, in part by the same epigenetic mechanisms that silence genes, presumably to keep our genomes from becoming completely dysfunctional as selfish DNA sequences battle to make as many copies of themselves as possible. It may come as no surprise that disruptions in the epigenetics of cancer cell genomes cause disruptions in the normal control of these mobile genetic elements, which may cause further genomic alterations as they copy themselves around the cancer genomes.

We don't yet know the extent to which these mobile elements inside the genome influence cancer susceptibility. But emerging evidence suggests that they may have a more important role than we formerly recognized. Several studies suggest that mobile elements can cause genome damage and "dysregulation of genome replication or cell cycling [and] disruption of cooperative cellular behaviour," and that disruption of normal expression in areas of the genome that harbor mobile genetic elements is prevalent across many cancers.

Given what we know about how the hallmarks of cancer map onto cheating in multicellular cooperation, we would expect that some aspects of cancer are a result of cheating in genomic cooperation—not just cheating in cell-level cooperation. Perhaps some cancers are driven by cheating DNA, whereas others are driven more by cellular cheating. As we learn more about cheating DNA, and more studies are designed to look specifically for it, we will be better able to answer the question of what role it might play in cancer.

For example, DNA replicating outside of the chromosomes could be a logical candidate for cheating DNA that could contribute to cancer. If the DNA is outside the chromosomes, its extrachromosomal existence implies that it has already escaped from genome-level controls of DNA replication and may be "free to proliferate or indulge in selfish behaviour," according to Featherston and Durand. But most pathology techniques and genomic sequencing methods used today don't even allow us to see whether there are any DNA sequences outside of the chromosomes. These methods either ignore extrachromosomal DNA sequences (DNA that is free floating in the nucleus of the cell) or wrongly assume they are part of a chromosome. Since we can't see this extrachomosomal DNA in standard sequencing, we can't see whether it is associated with cancer or cancer progression.

One exception is a study conducted by Paul Mischel and his team at the University of California, San Diego. Mischel is a kind and open-minded physician-scientist who isn't afraid to apply his considerable intellect to challenging the status quo. He and his team were looking at the role of extrachromosomal DNA in glioblastoma, a type of brain cancer. I was lucky enough to hear Mischel speak about his findings at a cancer evolution workshop, where he pointed out little specks of extrachromosomal DNA that he found in his samples. He discovered this extrachromosomal DNA in about half of his brain cancer samples and almost never in normal cells. And this extrachromosomal DNA contained extra copies of driver oncogenes (genes associated with cancer), suggesting that they may be playing a causal role in cancer rather than just being associated with the disease.

After his presentation, Mischel and I spoke about the possibility that these extrachromosomal DNA sequences may be selfish genetic elements, and if so, what this meant for understanding cancer evolution more generally. In my mind, this is one of the most exciting open questions about cancer evolution. The theoretical foundations of much of the field of cancer evolution would have to be reconsidered if cancer is, in part, a result of selection at the gene level favoring selfish genetic elements that can propagate themselves. If selfish genetic elements play a role in cancer, it would also be necessary to rethink our tools and methods so that we can detect and measure them. Our default assumption is that cancer is a result of evolutionary advantages that come to cells when they break free of their vehicles—multicellular bodies—and proliferate without the usual checks and balances that the organism imposes on it. But Mischel's work suggests that we might have been too quick in dismissing the possibility that cancer might come from *genes* that break free of *their* vehicles—the chromosomes—and replicate outside the usual constraints of the DNA replication process.

Cheating may not just be the domain of cancer cells but also a strategy used by the genes inside them. Similarly, we have seen in this chapter how cooperation is not just the domain of normal cells of the body but may also be a strategy used by cancer cells to better exploit the body.

Looking at cancer in the context of the ecology of the body, it is clear that cancer cell cooperation can contribute to cancer progression. This is ironic, given that cancer is fundamentally a problem of cheating in multicellular cooperation. But it's as though cheating only gets cancer cells so far—cooperation may be a strategy that allows them to quite literally get farther, allowing them to more successfully leave a primary tumor, invade new tissues, and metastasize. Cooperation can also enable cancer cells to accomplish things they could never do as single cells, like divide labor, break through membranes and tissues of the body, and survive in challenging conditions. Indeed, cancer cells cooperating may be even more dangerous to us than cancer cells cheating. Disrupting cooperation among cancer cells may be a critical tool in cancer treatment, especially during advanced stages of cancer when cancer cell cooperation is more likely to have evolved.

Evolutionary approaches—including disrupting cooperation among cancer cells—can help us do a better job of controlling cancer in the clinic. In the next chapter, we will see how an evolutionary and ecological approach offers us an opportunity to take the reins on cancer.

7

How to Control Cancer

Human history is a story of expanding control over our world. We build shelters to keep out the elements, we create infrastructure to deliver energy and water, and we cultivate plants and animals so that we have a steady food supply. Yet much of the world inside of us remains out of our control.

The body seems to rebel against itself as cancer progresses. All the controls the body has evolved to regulate cell cycles, cell metabolism, cell movement—they can all break or fail. When we treat cancer, we are trying to regain control of the internal world of the patient. But this can be much harder than it seems; our internal worlds are complicated ecological environments containing evolving populations of cells, and cancer itself keeps evolving, even in response to our treatments. However, thinking about the evolutionary and ecological nature of cancer as we are treating cancer and designing new therapies can help us better understand—and perhaps better control—cancer.

In 1972, just one year after signing the National Cancer Act, Richard Nixon signed an act creating a new national policy around an agricultural approach called "integrated pest management." In prior decades, farmers sprayed crops with chemical agents like dichloro-diphenyl-trichloroethane (DDT) to control insect populations and keep their fields pest-free. But heavy spraying of DDT had unforeseen consequences on ecosystems and human health, including the decline of bird populations and increased risk of cancers in humans.

The passage of this new national policy came on the heels of Rachel Carson's influential book *Silent Spring*, as well as growing public awareness of the dangers of pesticides. Chemicals like DDT were not only damaging to the environment and human health; we realized that they were unsustainable in the long run. Pests evolve resistance to the chemicals, and eventually, those chemicals become ineffective. In the case of DDT, selection favored pests that had mutations that changed the regulation of their sodium channels, which made them resistant to DDT. Pests that could pump out DDT by upregulating detoxification pathways could avoid the harmful effects of DDT.

Integrated pest management is a way of controlling agricultural pests by taking a long-term perspective that aims to avoid the evolution of resistance to chemical pesticides. One of the keys to the effectiveness of pest management is the idea that resistance to chemical pesticides has a cost for the organism doing the resisting, and so in the absence of the chemical, resistant individuals will actually be at a disadvantage. Thus, the first strategy of integrated pest management is to do nothing—to act only if the damage from pests reaches a critical threshold. The next strategy is to reduce the numbers of those pests, applying chemical treatment to bring them back below the threshold where they are not doing too much damage. Integrated pest management assumes that resistance is already present in the population. It is understood that if you apply too high a dose or spray your field too often, you will get rid of all the pests that are sensitive to this treatment and leave only those that are resistant to it, making it impossible to control pests in the long term. Integrated pest management anticipates this eventuality, using lower doses that do not get rid of all the sensitive pests but allow long-term control of the pest population.

The logic of integrated pest management was the inspiration for Bob Gatenby, a radiation oncologist and cancer researcher at the

Moffitt Cancer Center in Tampa, Florida, to develop a new approach to cancer therapy. When Gatenby learned about integrated pest management's strategy of leaving pests alone to prevent the evolution of resistance, he wondered if this kind of approach could be applied to cancer treatment. Gatenby started exploring these ideas in 2008, used his personal funds to run the initial preclinical study at the University of Arizona (where he was the chair of the radiology department at the time), and has been working ever since to apply these ideas from pest management to cancer treatment (now with grant funding from the National Cancer Institute and other organizations).

Just as we discovered in the case of pesticides, the biggest problem in cancer therapy is the evolution of resistance. During treatment, cancer cells evolve so that they are no longer sensitive to the therapy and the therapy stops working. The evolution of resistance to chemotherapy has been a problem for every kind of drug that has ever been tried, including targeted therapies like epidermal growth factor receptor (EGFR) blockades and human epidermal growth factor receptor 2 (HER2)-targeted therapies. Gatenby and his colleagues developed a revolutionary approach to cancer treatment in which the goal is not eradication of the tumor, but rather long-term control of the tumor. As with integrated pest management, this approach aims to limit the burden of the tumor while keeping cancer cells sensitive to therapy, in order to allow the same drug to be used indefinitely and limit damage to the environment (in other words, the patient).

Gatenby's approach is called adaptive therapy, to capture the idea that the therapy itself changes in response to the tumor. In adaptive therapy, a tumor is closely monitored through imaging or blood tests. The information about how a tumor is or isn't growing is then used to decide on the appropriate dose. There are several different algorithms for how to take this information about tumor growth and turn it into a dose decision, but the general principle is this: the algorithm is designed to find a dose level that will keep the tumor stable and at an acceptable size that is not too damaging to the patient. This is essentially integrated pest management for tumors.

The exact algorithm for adaptive therapy varies somewhat from study to study, but the process has the same goal of keeping the tumor stable and under control. First, the tumor is given a relatively high dose of the drug to make it smaller. (This also reduces the cancerous cells' population size, slowing the subsequent rate of evolution in the tumor.) Next, the tumor is regularly monitored and is treated with cancer drugs based on its behavior. If the tumor remains the same size, keep the dose the same. If it grows, increase the dose (up to some maximum tolerated dose). If it doesn't grow, back off by cutting the dose. If the tumor ever shrinks below some lower threshold, dosing is paused until the tumor grows back above that threshold. An alternative method is to keep the dose constant, but to pause the administration of the drug once the tumor is half of its initial size.

Adaptive therapy turns the framing on cancer upside down— rather than trying to destroy it, adaptive therapy allows the tumor to exist, but shapes it into something more controllable. It transforms cancer from an acute, lethal disease, into a chronic, manageable disease. Giving no doses or low doses of the drug allows the body to retain sensitive, less aggressive cells, making it possible to keep treating the tumor with the same drug. Since the intensity of the treatment increases only when the tumor is growing, selection may also disfavor fast-dividing cells, and favor cells that divide more slowly. This may slow the speed of cellular evolution inside the tumor as well. Adaptive therapy might not be the best option when it is clear that high-dose therapies can cure a cancer (for example, genetically homogenous tumors that are found early). But for more advanced cancers that are hard to control with traditional therapy, adaptive therapy offers an alternative to high-dose therapy that—as we will see later—has been successful in controlling late-stage cancer.

In a study published in 2009—the study that Gatenby ran using his own personal funds—he and his colleagues tested the adaptive therapy approach using a xenograft model, transplanting human ovarian cancer cells into mice. The mice were treated according to standard therapy protocols with a standard chemotherapy drug (three high-dose treatments in rapid succession) or adaptive therapy protocols. They also had a control condition in which the mice received no treatment. For the mice in the standard treatment condition, the tumors shrank at the beginning, only to rebound after a few weeks. The mice in the adaptive therapy condition, on the other hand, had tumors that stayed relatively stable throughout the experiment. They replicated this experiment and found the same results: adaptive therapy kept the mouse tumors under control.

Gatenby and his colleagues concluded that the adaptive therapy approach allowed these mice to "survive indefinitely with a small, reasonably stable tumor burden." In two other experiments, the team tested adaptive dosing algorithms with breast cancer cells from humans transplanted into mice. In both experiments, they found that the tumors could be controlled with a smaller and smaller dose as time went on. They also found that the tumors being treated with adaptive therapy actually displayed less necrosis (dead tissue) and a more stable blood supply, suggesting that adaptive therapy may actually help stabilize the resources and hazards in the ecological environment.

As we saw in the last chapter, more stable environments can select for cells that have slower life history strategies. More stable environments might therefore select for cells that are less aggressive and might also reduce selection for cancer cell cooperation. In the last chapter we also saw how unstable environments can create conditions that select for cooperation among cancer cells. Perhaps part of the success of adaptive therapy is a result of normalizing the resource flow and changing the selection pressures on cancer cells so that they don't evolve to cooperate.

After these experimental successes with adaptive therapy, Gatenby decided that it was time to test adaptive therapy in humans. Adaptive therapy is an example of a new kind of personalized therapy—one that is inspired by an understanding of the evolutionary dynamics of resistance. Adaptive therapy is not just personal, it is also dynamic, with dosing based on the growth of cancer cells and the response of the particular patient's tumor to therapy. It can also be used with any existing drug or treatment, lowering the barriers to clinical use. In addition, adaptive therapy can be used with any method of assessing tumor burden, from imaging of tumor size to measurement of prostate-specific antigen (PSA), a tumor marker for prostate cancer, in the blood.

In 2016, Gatenby teamed up with oncologist Jingsong Zhang to conduct the first clinical trial using adaptive therapy. They enrolled eleven men with metastatic prostate cancer that was no longer responding to hormone therapy in a pilot clinical trial. Prostate cancer cells normally need testosterone to proliferate, so hormone therapy suppresses testosterone to prevent those cells from spreading. But prostate cancer cells can become "castration resistant," often by producing their own testosterone. The drug abiraterone interferes with the synthesis of testosterone, so it's often prescribed for castration-resistant prostate cancer treatment—until those cells evolve resistance to abiraterone. The amount of time it takes to evolve resistance to abiraterone once treatment begins varies a lot from individual to individual. With normal continuous treatment, after 16.5 months, about half of men have tumors that progress (16.5 months is the median time to progression in general; this study did not include a control condition). In Gatenby's adaptive therapy trial, they measured tumor burden using PSA. They stopped dosing with abiraterone when PSA dropped below 50 percent of the starting level—leaving the tumor alone when PSA levels were low. They resumed treating with abiraterone only when PSA levels rose above 100 percent of the starting level. With adaptive therapy, Gatenby kept the tumors under control for much longer than standard therapy. As of October 2017, when Zhang and Gatenby's pilot study was accepted for publication, only one of the eleven patients' cancers had progressed. This is quite an astounding result: the median time to progression for the patients in the adaptive trial was at least 27 months, vastly longer than the typical 16.5 months. In fact, the median time to progression was probably much longer than 27 months (because so few patients progressed during the study it is impossible to calculate the actual median time to progression). In addition, the men on adaptive therapy had received less than half the total dose of abiraterone compared to the standard of care recommendation.

Gatenby is working on opening up clinical trials for adaptive therapy for melanoma, thyroid cancer, and ovarian cancer at Moffitt, and researchers at other institutions—including our team at ASU and the Mayo Clinic in Arizona—are starting clinical trials in other cancers using adaptive therapy. Adaptive therapy holds the promise of controlling cancer better and longer, as well as reducing the dosage required for keeping tumors under control. We are also collecting data on patient quality of life to formally test whether patients experience higher quality of life with adaptive therapy.

Phoenix from the Flames

As the story goes, the mythological phoenix burns in a great fire. And then, from the ashes arises a new, stronger, and more youthful bird. The phoenix, brilliant in red and gold, represents resilience and survival. It is also a fitting, if unfortunate, symbol of the resilience of cancer in the face of many of our treatments.

Like the phoenix, cancer is capable of rising from the ashes, gaining strength from the very forces that we would expect to destroy it. Cancer's resilience comes from its evolutionary nature: it is a population of diverse cells evolving rapidly under selective pressures. And when we treat cancer with radiation or chemotherapy, these therapies themselves become the selective pressures, selecting for cells that can survive the therapies. The next generation of cells in the tumor is made up of the descendants of those cells most capable of withstanding the therapy. So if therapy does not completely eliminate the cancer cells (which is often the case), the cancer may grow back. It also means that humans and our treatments of cancer are part of the selection pressures on the disease. We sculpt the evolution of tumors, whether we intend to or not.

We have made tremendous progress on how we treat cancer over the last few decades—some cancers, like thyroid cancer and childhood leukemia, can be cured. Some five-year survival rates are extremely high, near 100 percent for early-stage thyroid cancer and between 60 and 85 percent for childhood leukemias, depending on the type, according to the American Cancer Society. But we have also stalled out when it comes to our strategies for dealing with advanced stages of cancer. Metastatic cancer is a beast that we don't understand—and drug-resistant metastatic cancer is even more elusive. Our treatments for metastatic cancer do little more than extend life for a few months, and some studies suggest that patients can do just as well with palliative care (which is focused on improving patient quality of life and reducing pain) as with expensive and painful treatment that is aimed at curing the cancer.

This raises a question that evolutionary biologist and infectious disease expert Andrew Read has posed: "How should we treat patients when resistance to medications is already present in a tumor or infection and there are no other options?" Should we still use aggressive therapy, or should we consider a less aggressive approach that doesn't put such strong selection pressures on cells to evolve resistance? "Drug use causes drug resistance, a firestorm of drugs removes the competitors of the very things we fear: the cells and bugs we can't kill," notes Read. In other words, our therapies shape the evolutionary process happening inside tumors. And if we can't eliminate all cancer cells with aggressive treatment, we can end up inadvertently creating conditions that favor the cells we can't control.

My colleagues and I have wondered whether there is a way to tell the difference between those tumors that we can successfully treat with aggressive therapy and those that will evolve their way around our strategies. To address this question, we held a consensus meeting at the Wellcome Trust Genome Center. We established a set of principles for how to measure cancer's evolvability, called the Evo-Index and Eco-Index (figure 7.1). "One of the most important things to get about cancer is that it is constantly changing," said Carlo Maley, who led our consensus meeting, "Until now we haven't had a way to measure (or infer) the tumor's dynamics." The Evo-Eco Indexes provide guidelines for developing biomarkers to group cancers into different types based on their evolutionary and ecological characteristics.

Biomarkers help us diagnose cancer and employ risk stratification—helping us to predict which patients are likely to progress to metastatic cancer. Most biomarkers for cancer are molecular signatures, like a mutation in a particular gene or the presence of a particular

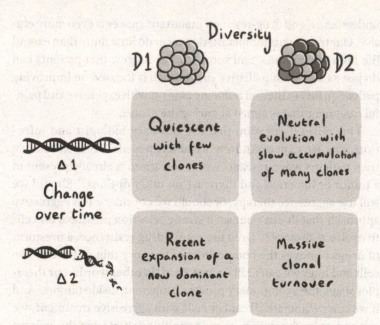

FIGURE 7.1 The Evo-Eco Indexes are a way of quantifying the evolutionary and ecological dynamics of tumors that can help us to classify and more effectively treat cancer. The Evo-Index (shown here) has two dimensions: diversity of the tumor and genetic change in the tumor over time. Tumors can be high or low in both of these dimensions, leading to a 2 × 2 classification scheme of tumors into four different types. These different types of tumors are likely to respond differently to treatments, and so this typology may be an effective guide for personalized cancer treatments. The Eco-Index (not shown) has two dimensions as well; availability of resources and presence of hazards, forming a complementary 2 × 2 classification scheme. Together the Evo and Eco-Indexes may help us classify and more effectively treat cancer.

receptor. Biomarkers are usually specific to a certain type of cancer or provide information about likely responses to a particular therapy.

Standard methods in pathology typically consist of taking a biopsy of a tumor and looking at it under a microscope, and looking at the genetics of the tumor at a single time point. Cancer biologists sometimes say that traditional pathology is like trying to infer the rules of football by looking at pictures in *Sports Illustrated*—you see snapshots and none of the action.

What the evolutionary and ecological framework offers is a new approach to personalized medicine in cancer. Rather than looking for biomarkers like a particular mutation and treating based on that mutation, the Evo-Eco Indexes are ways of classifying (and eventually treating) tumors based on their evolutionary and ecological features. "What we are trying to do with the Evo- and Eco-Indexes is learn the rules of the game by measuring the dynamics, incentives and pressures on tumor cells," said Maley. The goal of the Evo-Eco Indexes is to measure the key evolutionary and ecological parameters in a tumor to better predict how it will evolve.

The Evo-Index is made up of two components: the genetic diversity of the tumor and the rate of genetic change in the tumor. The Eco-Index has two components as well: the resources (like blood supply) and hazards (like immune predation) present in the environment around the cells. Together, these factors can help to predict how tumors are likely to evolve, and, critically, how they are likely to respond to different types of therapies.

The Eco- and Evo-Indexes are frameworks for helping us identify tumors that are likely to have different underlying evolutionary dynamics. The hope is that by measuring these ecological and evolutionary dynamics, we can design and modulate our therapies based on the particular dynamics of the tumor, as opposed to just treating based on the static properties of a tumor, like its genetic profile from a single time point. Not all tumors are the same, and the goal of the Evo-Eco Indexes is to identify the evolutionary and ecological parameters that are most important for shaping the tumor dynamics and then use this knowledge to shape tumor evolution in the direction we want it to go—rather than just letting evolution take its course.

We have seen that cancer is a complex, adaptive enemy. It is a population of cells that evolve in response to whatever we throw at them. And so if we want to control cancer we may need to get smarter and more strategic about how we approach it. Most traditional therapies adopt a machine-gun approach, aggressively blasting a tumor with destruction and chaos. But evolutionarily informed therapies, such as adaptive therapy, use the tactic of shaping the cells' fitness landscape to change the evolutionary trajectory of the tumor. There are many potential strategies that we can use to control populations of cancer cells by shaping the way they evolve: slowing them down, redirecting them, and culling only those that are most likely to be a problem.

One way that evolutionary biologists think about evolving populations is using the metaphor of a fitness landscape. The tops of hills are peaks in fitness and valleys represent low fitness. As populations evolve, they are described as climbing hills in their fitness landscape (figure 7.2), reaching points of higher fitness as they mutate. One way to control cancer would be to manipulate this fitness landscape—to alter the peaks and valleys—so that cancer cells evolve in the direction we want them to—for example, trapping them on "local peaks" so they remain relatively harmless. If we can shape this fitness landscape, we can shape how cancer cells evolve. We can shape them to be less resistant to our therapies. We can shape them to be less aggressive. We can shape them to stay put rather than invading and metastasizing. We may even be able to send them down blind alleys or—in the language of fitness landscapes—send them up local maxima so they don't evolve to threaten our survival and well-being (like sending them up a small hill so they don't find the mountaintop). We just need to figure out how to mold the fitness landscape into the right shape so we can get cancer exactly where we want it—or at least keep it from evolving up to a fitness peak that could kill us.

Earlier in this chapter I talked about the early successes and promise of adaptive therapy, which adjusts the dose based on tumor growth. What other strategies can be used for controlling cancer? Can we use evolutionary and ecological principles to devise new strategies for controlling cancer rather than eradicating it?

SLOW IT DOWN

The simplest strategy for controlling cancer is to slow the rate of cellular evolution, making it take longer for cells to cross over the fitness landscape. If, for example, we can slow it down by half, it would take twice as long for a growth to become cancerous. Since it often

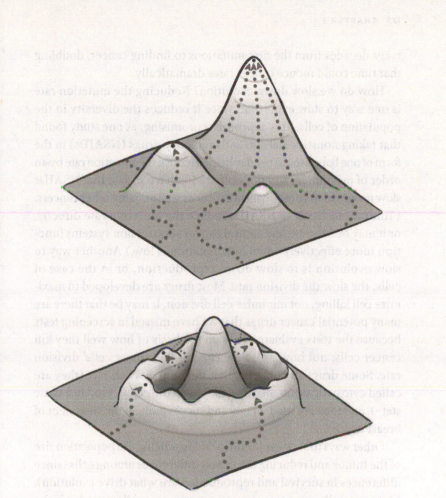

FIGURE 7.2 One of the reasons that cancer evolves in the body is because cells with cancerous properties have higher fitness, represented here by the highest point on the fitness landscape. Fitness landscapes are a way of representing the trajectory of evolution of a population. As cells mutate in the body, they acquire mutations—some of which move them to a higher point in the fitness landscape. Those cells with higher fitness will be more likely to leave progeny, moving the population of cells up the fitness landscape closer to the cancerous peak (*top*). One way to prevent cancer is to shape the trajectory of cells on the fitness landscape, or even to modify the fitness landscape itself, for example, by changing the environment of the body (*bottom*). For example, it may be possible to trap cancer cells on "local peaks" that essentially restrict cells to local maximum rather than allowing them to evolve to the global maximum—the highest point on the landscape. Manipulating the fitness landscape to shape the trajectory of cancer evolution is one of the strategies for evolutionarily inspired cancer control.

takes decades from the first mutations to finding cancer, doubling that time could reduce cancer rates dramatically.

How do we slow down evolution? Reducing the mutation rate is one way to slow evolution, since it reduces the diversity in the population of cells. This approach is promising, as one study found that taking nonsteroidal anti-inflammatory drugs (NSAIDs) in the form of one baby aspirin per day helped reduce the mutation rate by an order of magnitude, and multiple studies have shown that NSAIDs slow progression to esophageal cancer as well as many other cancers. (This may be because NSAIDs reduce the mutation rate directly, or it may be because our natural cancer suppression systems function more effectively when inflammation is low.) Another way to slow evolution is to slow down reproduction, or in the case of cells, the slow the division rate. Most drugs are developed to maximize cell killing, not minimize cell division. It may be that there are many potential cancer drugs that we have missed in screening tests because the tests evaluate drugs on the basis of how well they kill cancer cells, not how well they control the cancer cells' division rate. Some drugs already exist that slow the division rate (they are called cytostatic drugs, because they keep the cell, "cyto," in a static state), and they are used widely and successfully in the treatment of breast cancer.

Other ways to slow evolution include reducing the population size of the tumor and reducing the fitness differences among cells (since differences in survival and reproduction are what drive evolution). More generally, we can try to encourage cancer cells to evolve in the direction of a slower life history strategy by shaping the ecological environment they are in. Slowing down the cell strategy means that cancer becomes the dormant, quiet roommate we met in chapter 3.

FAKE THE DRUGS

Gatenby's arsenal of evolutionarily inspired strategies for cancer control has a lot of clever and creative ideas for how to keep cancer cells from gaining the upper hand. Building on the notion that resistance is costly—that cells have to work and expend energy to be resistant to drugs—Gatenby had the idea that he could make the resistant cells expend this cost without necessarily getting a benefit. Cells that are resistant to multiple drugs often have efflux pumps that require energy to run—and these pumps get rid of the drugs inside the cell.

Gatenby thought this resistance to multiple drugs may actually be a vulnerability that could be exploited—by giving the cells "fake drugs," in other words, nontoxic or minimally toxic substances. These fake drugs activate the efflux pumps of cancer cells and cause them to expend energy, without actually giving them a survival benefit over nonresistant cells. Gatenby calls these drugs "ersatzdroges"—because it sounds better than "fake drugs" and means the same thing. (*Ersatz* means "substitute" in German.) Gatenby and his colleagues found that they could decrease cell proliferation of resistant cells in Petri dishes and that the growth rate of resistant cell lines (compared to similar nonresistant cell lines) was lower in a mouse model with the administration of the ersatzdroges. This strategy makes resistant cells work hard without any gain—they run their molecular motors to pump out substances that are not actually drugs—and have fewer resources left for proliferation as a result.

Get Back to Basics

Gatenby has a knack for developing novel strategies for cancer control. We know that cancer transforms the ecology of the body, making the tumor microenvironment more acidic. This acid contributes to breaking down the extracellular matrix, which not only destroys the internal environment of the body—it also paves the way for invasion and metastasis, making it easier for cells to leave the degraded environments they find themselves in. Gatenby, armed with this knowledge, decided to do an experiment to see whether sodium bicarbonate (baking soda) could reduce metastases in mice.

Gatenby and his team gave mice baking soda orally after they had been given injections of breast cancer cells, prostate cancer cells, or melanoma cells. They found that giving mice baking soda reduced the acidity of the tumor environment and that it led to a "significant reductions in the number and size of metastases to lung, intestine, and diaphragm." The size of the primary tumor was not affected, but by returning the tumor environment to a more pH neutral state, the metastases were significantly decreased—and this led to an improvement in survival for the mice receiving what Gatenby's team called "bicarbonate therapy." Decreasing the acidity of the environment may be affecting the life history evolution of cancer cells inside the tumor as well—reducing the hazards that cells face from high acid and also reducing the opportunities for colonization—both factors that could select for a slower life history strategy for the cells.

Feed the Tumor

Hypoxia, or low levels of oxygen in a tumor, is a critical part of the tumor microenvironment. When oxygen levels are low, cancer cells are more likely to invade and metastasize. Low levels of resources select for cancer cells that move more readily. Studies suggest that normalizing the resource delivery to the tumor can actually reduce metastasis, and that using low levels of antiangiogenic drugs (which help to regulate blood flow to the tumor) can improve response to treatments. As we saw earlier in this chapter, more normalized resource flow is a consequence of adaptive therapy as well—which could be contributing to its success.

Normalizing the resource flow to the tumor is likely to have an impact on the life history selection pressures on cells in that tumor. In general, having access to stable but low levels of resources selects for individuals with slower life history strategies. In a tumor with normalized resources, this is likely to be the case as well; with normalized resources, selection favors cells with slower rates of proliferation and a lower tendency toward dispersal.

This strategy of providing stable resources for a tumor sounds counterintuitive—shouldn't we be trying to starve the tumor? The problem is, of course, that starving the tumor just makes the cells more likely to change their gene expression states to move, and it also selects for cells that are more likely to move. Feeding a tumor (at a stable and low level) might allow the tumor to keep growing where it is—but that is generally far preferable to encouraging it to invade and metastasize. It's similar to what Winston Churchill said about democracy: that it is the worst form of government except for all those other forms that have been tried. Feeding the tumor might seem like a bad strategy because it is allowing cancer cells to have access to resources that allow them to proliferate and grow where they are, but that may be much better than the alternative of encouraging metastasis.

Leveraging Cooperation Theory for Cancer Control

Another way to get control over cancer is to support our body's ability to effectively detect and respond to cellular cheating. Our body protects us by detecting cellular cheating and shutting that cheating down before it can evolve into cancer. If we can enhance our body's ability to detect cellular cheating, or restore these abilities if they get disrupted (by mutations in our cancer suppression systems, for example), we may be able to better prevent and treat cancer.

Our body's multilevel cheater detection systems are in place to detect cellular cheating and protect us from the possibility that those cellular cheaters will initiate cancer or urge on its progress. But unfortunately our body's cheater detection systems are not immune to tampering. And cancer is under constant selective pressure to trick and trip up our cellular cheater detection systems. By the time cancer has grown large enough to be diagnosed, it has evolved to evade and hijack these cellular cheater detection systems on every level. Many cancers have cells with mutations in *TP53*, the information processing hub for the cell's intrinsic cheater detection system. The neighborhood detection systems is disrupted as well—often cancer cells produce factors that trick their neighbors into tolerating proliferation and cell movement as "normal" behaviors. And of course, cancer evolves to evade the immune system. Like a population of prey evolving to get better at running and hiding from predators, cancer cells evolve to stay under the radar of the immune system.

When cancer disrupts our cellular cheater detection systems, what can we do to get cancer under control?

One possibility is to reboot cellular self-control, getting our intrinsic cancer suppression systems functioning properly again. Each of our cells has within it a complex genetic network that allows it to monitor its own behavior and adjust its gene expression and behavior if it detects that something is wrong. For example, the genetic network around *TP53* will halt the cell cycle, start DNA repair, and induce cell suicide if necessary. In cancer this cell-intrinsic cheater detection system is often lost when *TP53* gets mutated or deleted. Some potential strategies for rebooting cellular self-control are restoring *TP53* function when it is lost, or simply promoting cellular behaviors such as DNA repair. Many therapies have been aimed at trying to reactivate apoptosis pathways in cells that are surviving when they shouldn't, but these are notoriously difficult to use in the long term because they rapidly select for cells that are resistant.

Lisa Abegglen and Joshua Schiffman, the cancer biologists at the Huntsman Cancer Institute whom we met earlier in this book, are now working on developing new therapeutics using elephant *TP53*. They have shown that elephant *TP53* can restore normal p53 function and apoptosis in human osteoscarcoma cells and are now conducting experiments with mice to test whether elephant *TP53* can induce apoptosis in mouse tumors in vivo.

We are just beginning to understand how our cells process and integrate information through genetic networks like those around *TP53*. As we come to understand how cells leverage information to help keep us cancer-free, we may be able to leverage this for better cancer prevention. We can also look to the collective intelligence of our bodies—and the bodies of particularly cancer resistant organisms like elephants—for inspiration about how to do a better job at controlling cancer in the clinic. These approaches suggest that we should be integrating many sources of information for measuring and monitoring cancer and developing decision-making tools for using that information most effectively.

As we have seen, multicellular bodies evolved to make it very difficult for cells to cheat and get away with it, and we have a whole toolbox full of cancer suppression systems that help to police and control cellular cheating. For example, *TP53* only allows normally behaving cells to divide, and *BRCA* repairs broken DNA that could contribute to cellular cheating. These cancer suppression mechanisms cannot completely eliminate cellular cheating, but they can go a long way toward keeping this cheating under control and allowing the multicellular organism to live a long and healthy life. In other words, multicellular life has largely solved the problem of cellular cheating by evolving an arsenal of cancer suppression mechanisms including cancer suppressor genes like *TP53*, DNA repair systems, and perhaps even the ability to generate noncancerous clonal expansions to prevent cancerous ones from growing.

A better understanding of how our bodies suppress cancer—and how other organisms across the tree of life suppress cancer—can point the way to new solutions for controlling cancer and extending life. We may discover new cancer suppression systems—like the kind of cell extrusion that placozoa may engage in when cells mutate too extensively—or figure out ways to better leverage known cancer suppression systems like *TP53* by looking to large animals like elephants to understand how they use these systems to suppress cancer. We can also use an evolutionary perspective to better understand the nature of the trade-offs that come with some forms of cancer suppression—why, for example, mutations in the cancer suppressor gene *BRCA* are associated with both higher rates of breast cancer and higher fertility.

Another way to enhance the body's cheater detection systems is to get the neighborhood-level monitoring systems back on track by essentially encouraging cells to spy on their neighbors. Often, cancer cells will hijack the normal cells around them by producing wound healing factors. These signals essentially tell the entire cellular neighborhood to tolerate behaviors including proliferation and cell movement (i.e., they induce neighbors to raise their threshold for detecting cheating). Wound healing signals allow cancer cells to engage in problematic behaviors without the normal checks and balances from their cellular neighbors. This may be one reason why reducing inflammation helps reduce the risk of cancer (lowering inflammation also reduces reactive oxygen species that can directly cause DNA mutations and small deletions). Reducing inflammation might also help to clean up the signaling environment and allow normal cells to properly detect cancer-like behavior of problematic cells in their neighborhood. Without all the "noise" of inflammation, it may be easier for our immune cells to focus in on the "signal" (i.e., cancer cells).

We can also get cancer under control by re-engaging the immune system so that it can keep cancer at bay. As we have seen, cancerous cells evolve many strategies to avoid being detected by the immune system. But it is possible to retrain the immune system to respond to cancer and even keep cancer cells from hiding from the immune system. These are the goals of immunotherapy. Sometimes cancer cells hide from immune cells directly by changing the proteins on their surfaces so they look normal to immune cells, and other times they hijack immune cells to signal for them that everything is normal and they should be left alone. Cancer cells can also directly interfere with the cheater detection systems that immune cells are equipped with.

Normally, our immune system works through a system of checks and balances so that it can react to threats (like cancer cells and disease-causing microbes) but also de-escalate when the threat has passed. Part of how our immune system does this is through immune checkpoints that halt the immune response if they receive information that there is no threat. Immune checkpoints are essentially systems that detect that the environment is cheater-free—and tell the immune system to stand down. This ability to de-escalate the immune response is critical to our health—if it were not in place, we would suffer from autoimmunity and excessive inflammation. But this also introduces a vulnerability: cancer cells can evolve to produce factors that trip these immune checkpoints and call off the immune response.

Some of the most promising approaches in immunotherapy are those that interfere with this ability of cancer cells to de-escalate the immune response. Immune checkpoint blockade therapies block the molecules that cancer cells produce that would otherwise deactivate the immune response. By restoring the immune system's ability to detect cellular cheaters, immune checkpoint blockade therapies have been successful at treating previously intractable cancers, including melanomas and lung cancers, in some patients.

Despite some early setbacks, immunotherapy is now one of the most promising frontiers of cancer treatment. Nevertheless, cancer cells do still evolve resistance to immunotherapies just as they evolve resistance to traditional therapies. This means that evolutionary management is important for treatment with immunotherapies as well.

Cooperation, Interrupted

We saw in the last chapter that cancer cells not only evolve to cheat in the foundations of multicellular cooperation, but they can cooperate with one another to better exploit their host. As disturbing as this is, it suggests another potential target for cancer control. It may be possible to disrupt cancer cells' cooperation with one another. For example, interfering with adhesion in circulating cell clusters could possibly reduce the likelihood of metastasis (since cell clusters have been found to be more likely to metastasize than single cells). Disruption to cooperative cellular signaling among cancer cells is another potential strategy for cancer control.

Our bodies most likely evolved to suppress cooperation among those misbehaving cells that shouldn't be cooperating. Think of it this way: evolution has shaped our cancer suppression systems to keep cancer cells from cooperating with each other. This suggests important directions for cancer treatment. Can we perhaps disrupt cancer cell cooperation to prevent cancer from evolving to advanced metastatic stages? And can we disrupt the cooperation and coordination in cancer cell colonies to better treat metastatic cancer?

I shared these ideas and questions with Gunther Jansen, an infectious disease researcher who, like Andrew Read, was taking on a lot of the traditional assumptions about how to best control infectious disease. We were both intrigued by the idea that we could apply strategies from infectious disease control to cancer control, and, specifically, how we could find ways to disrupt communication and cooperation between cheating cells.

Quorum quenchers are biological tools used in infectious disease control; these molecules block other molecules that bacteria use for communication and cooperation. They have been successfully deployed to reduce bacterial population sizes, and prevent bacteria from forming structures like biofilms that contribute to drug resistance. There are drugs on the market that target cancer cell communication molecules, like growth factors, angiogenic factors, and immune suppression factors. But cancer drugs are typically chosen by drug developers based on their ability to destroy cancer cells. Perhaps we should be searching for and using drugs that disrupt cell cooperation by interfering with cancer cell communication even if they do not result in the eradication of the cancer cells.

We should also consider developing drugs that interfere with cancer cell aggregation. If we make it more difficult for cancer cells to stick together when they are circulating in the bloodstream, for example, we might find that this reduces metastasis. As we saw in the previous chapter, circulating clusters of tumor cells are more likely to successfully metastasize. These circulating clusters of tumor cells stick together using adhesive molecules called plakoglobins. Higher levels of plakoglobins are associated with worse patient outcomes, suggesting that they might be a potential candidate for therapeutic interventions. If we can interfere with tumor cells and make it more difficult for them to stick together in the bloodstream, we might be able to decrease the chances of metastasis.

In addition to cooperating with one another, cancer cells can hijack the cooperative nature of our normal cells. Cancer cells can send signals to normal cells requesting more resources, protection, and other benefits. The normal cells of our body are primed to cooperate because that's what cells in multicellular bodies do. As multicellular organisms, we have been selected to have cells that cooperate with one another to make us functional organisms. And part of that cooperation is a willingness to help and respond to signals from other cells. Cancer cells can hijack these cooperative signaling systems, exploiting the fundamentally cooperative nature of the multicellular body. They can also hijack the immune response and co-opt support cells to create an environment that helps them survive and proliferate. When cancer cells co-opt the body's immune and support cells to promote themselves, this creates an ecological niche in which cancer cells can thrive at the expense of the well-being of the multicellular body. This is why immunotherapy is an essential tool in the arsenal of evolutionarily informed cancer treatment.

As we come to understand cancer cell cooperation—both how it evolves and how it manifests in cell-to-cell interactions—we will be able to design interventions that are targeted at disrupting that cooperation. Interfering with public good production is one example of an intervention targeted at disrupting cancer cell cooperation. As we have seen, cancer cell cooperation may be particularly important during invasion and metastasis because colonizing new environments is challenging. If we can disrupt the cooperative interactions among cancer cells that help them to survive and thrive as they invade and colonize new tissues, we may be able to greatly reduce the burden of cancer. Similarly, if we can understand the evolutionary mechanisms that lead to the evolution of cooperation among cancer cells, we will have more tools for shaping the evolutionary trajectory of the cancer cells so that they don't evolve to cooperate at our expense.

Cure via Control

Will the future hold a cure for cancer? The answer to this question depends on how we define a cure. Does a cure mean a complete eradication of all cells in the tumor and any of their descendants? Or does a cure mean the ability to keep under control a tumor that is not growing, not invading, and not threatening the life of a patient?

We have seen that cancer is a result of evolution going on among cells in the body. An evolutionary approach tells us that being out of control is fundamentally part of cancer's nature. That is the essence of cancer: uncontrolled cellular behavior that doesn't follow the proper rules for multicellular cooperation. Cancer is a result of cell-level evolution happening within our bodies, favoring cells that don't do what they should do to help our bodies thrive. Under normal circumstances, our bodies are detecting and suppressing cells that are cheating in multicellular cooperation. But that cell-level evolution gets out of control with cancer, especially during advanced stages.

We also know that cancer continues to evolve, even (especially) when we treat it. So to cure an advanced cancer by eradicating it will be extremely difficult, because cancer can evolve in response to whatever treatments we throw at it. But there is another possibility— we can try to cure cancer by controlling it. This kind of cure via control is a distinct possibility, exactly *because* cancer can evolve in response to our treatments.

Cancer evolves, but we have the ability to anticipate that evolution and strategically plan our response. We can trick it, send it down a blind alley, sucker it into vulnerability, and shape it into something we can live with. If we can design our treatments to select for less aggressive cancer cells, support normal cells, and keep a tumor from getting too large—as with adaptive therapy—we may be able to get cancer under control and keep it under control.

Controlling cancer should be a goal for those cases in which a cure (i.e., complete remission with low rates of recurrence) is not likely with traditional therapy. If we continue to adopt the traditional notion of war when we think about curing cancer, then stopping curative treatment looks like "giving up" or "losing the battle." In a personal essay about his challenges with cancer, the late evolutionary biologist Stephen Jay Gould said, "I prefer the more martial view that death is the ultimate enemy—and I find nothing reproachable in those who rage mightily against the dying of the light." But war doesn't have to be about ammunition and aggression. It can also be about anticipating the moves of your enemies and outwitting them.

We have much to gain from accepting that cancer is a part of us and preparing for a long-term strategic interaction with it as an unpredictable and adaptive counterpart. It takes courage to face this truth and accept an uncertain future with cancer rather than hanging onto false hope that we will one day find a magical weapon that can target and eliminate cancer from the world.

Many researchers in the cancer community are working to identify the key parameters (for example, in the Evo-Eco indexes) that should be used when we're making strategic decisions about how to treat cancer with treatment approaches like adaptive therapy. The goal is to identify these parameters and use them to make wise choices regarding when to go with an aggressive treatment and eradication versus when to go with control and containment. This shift in our thinking allows us all to consider cancer as a chronic and manageable disease, which opens up new pathways for treating and preventing cancer.

So what should we do when our bodies fail to keep cancer under control? Ultimately, the right answer will depend on the kind of cancer a patient is facing, the stage of that cancer, and the preferences that the patient has for how they want to live the rest of their lives. If the cancer is highly treatable, and/or in an early stage, then aggressive treatment aimed at a cure may well be the right choice.

This brings us back to the different approaches of Athena and Ares that I mentioned at the opening of the book. Athena is the goddess of wisdom *and* war. Ares, the god of war only, lives for aggression for the sake of itself, thriving on the chaos of battle. Athena, on the other hand, uses her wisdom and strategic reasoning to keep the enemy at bay while avoiding the need for a costly battle. We need to control the enemy rather than leaving death and destruction in our wake, and causing massive collateral damage. Standard high-dose chemotherapy is the Ares approach to war. Adaptive therapy and some of the other strategies we talked about in this chapter are the Athena approaches to war.

Cancer affects us all. It is the second-leading cause of death globally. It affects our families, our communities, our world, and our worldview. Its reach is far greater than this: it goes beyond modern humans today, all the way back to the origins of multicellular life and across the tree of life. By looking at cancer from this broader evolutionary perspective, we can see that we are not alone in our struggle with cancer—life has struggled with it since the dawn of multicellularity. And we can see that evolution is both the reason we get cancer and our hope for keeping cancer under control.

We can shape a better future for human health and well-being if we understand and acknowledge our evolutionary past with cancer. It has been part of life since the earliest beginnings of multicellularity and it has evolved with us every step of the way. It has been a constant companion through our evolutionary history. We have lived with this freeloading roommate, this free rider, since our very beginnings. We have nevertheless managed to be evolutionarily successful, despite this unwelcome companion.

Evolution is a powerful force. It has shaped the diversity of life on our planet and the diversity and resilience of cancer cells inside our bodies. Our best hope for reducing the burden of cancer is to take this power into our own hands—by shaping tumor evolution along a trajectory that keeps it from evolving into something that can kill us and something we can't control. On one hand, cancer evolution is out-of-control evolution happening in our bodies. But on the other hand, we may be able to control the trajectory of that evolutionary process more than we realize—through measuring the dynamics of tumors and deploying therapies that shape the evolution of tumors in our preferred directions.

Evolutionary biology, ecology, and cooperation theory offer starting points for developing some of these tools. Approaches like the Evo-Eco Indexes—which quantify the evolutionary and ecological dynamics of tumors to help predict how tumors will respond to treatment—could help us distinguish between those cancers that are likely to be eradicated by high-dose therapy and those that might be better treated with approaches like adaptive therapy. Rather than fighting an all-out war on cancer, we can launch an intelligence-driven campaign—leveraging information from many sources to make smart decisions about how to control cancer and shape it into a companion that we can live with.

Cancer is a part of our past. And it will almost certainly be part of our future. But it is up to us what the future of cancer looks like, if it will continue to be an insurmountable enemy or whether we can transform it. We have an opportunity to leverage our collective human intelligence to do a better job of controlling cancer so that we can live longer and healthier lives. This opportunity requires interdisciplinary cooperation, effective communication, and an urgent sense of our shared goals—to find a way to better control cancer and support human flourishing. Cancer is a disease that defies all disciplinary boundaries. It invades every aspect of our lives and crosses into every territory we have settled. And so it will require a radically interdisciplinary and radically cooperative approach to understand its nature and how to most effectively treat it.

We may not be able to eliminate cancer, but we can create a world in which we redefine a cure as long-term cancer control. We can focus on preserving human life and quality of life as much as we can. That kind of future is within our reach.

Chapter 1. Introduction

2   *Remnants of cancers have been found in the skeletons of ancient humans, from Egyptian mummies* A. R. David and Michael R. Zimmerman, "Cancer: An Old Disease, a New Disease or Something in Between?," *Nature Reviews Cancer* 10, no. 10 (2010): 728–33.

2   *to Central and South American hunter-gatherers* Luigi L. Capasso, "Antiquity of Cancer," *International Journal of Cancer* 113, no. 1 (January 2005): 2–13.

2   *in 1.7-million-year-old bones of our early human ancestors in "the cradle of humankind" in South Africa* Edward J. Odes et al., "Earliest Hominin Cancer: 1.7-Million-Year-Old Osteosarcoma from Swartkrans Cave, South Africa," *South African Journal of Science* 112, no. 7/8 (July 2016), https://doi.org/10.17159/sajs.2016/20150471.

2   *from mammals, fish, and birds* Capasso, "Antiquity of Cancer," 2–13.

2   *Cancer goes back as far as the days when dinosaurs dominated life on our planet* Bruce M. Rothschild, Brian J. Witzke, and Israel Hershkovitz, "Metastatic Cancer in the Jurassic," *Lancet* 354, no. 9176 (July 1999): 398.

2   *Cancer began before most of life as we know it even existed* C. Athena Aktipis et al., "Cancer across the Tree of Life: Cooperation and Cheating in Multicellularity," *Philosophical Transactions of the Royal Society of London, Series B: Biological Sciences* 370, no. 1673 (2015), https://doi.org/10.1098/rstb.2014.0219.

4   *a crested saguaro cactus,* Carnegiea gigantea Lon & Queta, Crested Saguaro Cactus with Half Moon Behind; SE Arizona is licensed under CC-BY-NC-SA 2.0.

4   *a brain cactus,* Mammillaria elongata cristata David J. Stang, *Mammillaria elongata cristata* is licensed under CC BY SA 4.0.

4   *a "totem pole cactus,"* Pachycereus schottii f. monstrosus (Valentino Vallicelli, *Pachycereus schottii* f. *monstrosus* is available under CC-BY-SA.

4   *a* Cereus jamacaru f. cristatus Valentino Vallicelli, *Cereus jamacaru* f. *cristatus* hort is available under CC-BY-SA.

5   *from wild animals, to animals kept in zoos, to the domesticated animals that live with us in our own homes* Aktipis, "Cancer across the Tree of Life."

8   *"Nothing in biology makes sense except in the light of evolution"*   Theodosius
    Dobzhansky, "Nothing in Biology Makes Sense Except in the Light of
    Evolution," *American Biology Teacher* 35, no. 3 (March 1973): 125–29.

11 *they report being less likely to engage in some cancer-prevention behaviors,*
    like stopping smoking   David J. Hauser and Norbert Schwarz, "The War on
    Prevention: Bellicose Cancer Metaphors Hurt (Some) Prevention Intentions,"
    Personality and Social Psychology Bulletin 41, no. 1 (January 2015): 66–77.

11 *aggressive language related to treatment can increase stress levels for cancer*
    patients and their families   Aria Jones, "An Open Letter to People Who Use
    the 'Battle' Metaphor for Other People Who Have the Distinct Displeasure
    of Cancer," *McSweeney's Internet Tendency* (San Francisco: McSweeney's
    Publishing, October 19, 2012), https://www.mcsweeneys.net/articles/an
    -open-letter-to-people-who-use-the-battle-metaphor-for-other-people
    -who-have-the-distinct-displeasure-of-cancer; Katy Waldman, "We're
    Finally Winning the Battle against the Phrase 'Battle with Cancer,'" *Slate*,
    July 30, 2015, https://slate.com/human-interest/2015/07/how-battle-with
    -cancer-is-being-replaced-by-journey-with-cancer.html.

Chapter 2. Why Does Cancer Evolve?

14 *Peter Nowell described cancer as an evolutionary process based on the*
    accumulation of genetic mutations   Peter C. Nowell, "The Clonal Evolution
    of Tumor Cell Populations," *Science* 194, no. 4260 (1976): 23–28.

14 *John Cairns first pointed out that our bodies are likely to have protective*
    mechanisms that help to keep cancer from evolving within us   J. Cairns,
    "Mutation Selection and the Natural History of Cancer," *Nature* 255,
    no. 5505 (1975): 197–200.

14 *the idea of the stepwise progression of cancer (proposed by Leslie Foulds in*
    the mid-1900s)   Michel Morange, "What History Tells Us XXVIII. What
    Is Really New in the Current Evolutionary Theory of Cancer?," *Journal of*
    Biosciences 37, no. 4 (September 2012): 609–12.

14 *cancer biologist Mel Greaves*   M. F. Greaves, *Cancer: The Evolutionary*
    Legacy (Oxford: Oxford University Press, 2000).

14 *evolutionary geneticist Leonard Nunney*   Leonard Nunney, "Lineage
    Selection and the Evolution of Multistage Carcinogenesis," *Proceedings of*
    the Royal Society of London, Series B 266, no. 1418 (March 7, 1999): 493–98.

14 *computational evolutionary biologist Carlo Maley*   M. Greaves and C. C.
    Maley, "Clonal Evolution in Cancer," *Nature* 481 (2012): 306–13; Lauren F.
    Merlo et al., "Cancer as an Evolutionary and Ecological Process," *Nature*
    Reviews Cancer 6, no. 12 (2006): 924–35.

16 *having such extravagant sexual ornaments that the entire population becomes*
    catastrophically vulnerable to predation   Kalle Parvinen, "Evolutionary
    Suicide," *Acta Biotheoretica* 53, no. 3 (2005): 241–64.

21 *Sun Tzu warns against entering battle without knowing your enemy*   Sun Tzu,
    The Art of War: Complete Texts and Commentaries, trans. Denma Translation
    Group (Boulder, CO: Shambhala Classics, 2005).

23 *he described organisms as vehicles designed by natural selection to get genes into the next generation*   Dawkins, *The Selfish Gene* (Oxford University Press, 1976).

Chapter 3. Cheating in Multicellular Cooperation

27 *Talking about cancer cells as cheaters is shorthand for saying that they evolved to behave in an exploitative way*   Melanie Ghoul, Ashleigh S. Griffin, and Stuart A. West, "Toward an Evolutionary Definition of Cheating," *Evolution: International Journal of Organic Evolution* 68, no. 2 (February 2014): 318–31.

27 *My perspective on cancer is founded in the evolution of multicellularity and how cancer evolves as a "cheater" in the context of multicellular cooperation*   Aktipis, "Cancer across the Tree of Life"; Nunney, "Lineage Selection and the Evolution of Multistage Carcinogenesis."

29 *The hallmarks of cancer were laid out in a landmark paper by the cancer biologists Douglas Hanahan and Robert Weinberg in 2000*   D. Hanahan and R. A. Weinberg, "The Hallmarks of Cancer," *Cell* 100, no. 1 (2000): 57–70.

29 *updated a decade later to include two emerging hallmarks and two enabling characteristics*   Douglas Hanahan and Robert A. Weinberg, "Hallmarks of Cancer: The Next Generation," *Cell* 144, no. 5 (March 2011): 646–74.

29 *The hallmarks of cancer*   D. Hanahan and R. A. Weinberg, "The Hallmarks of Cancer." *Cell* 100 (2000): 57–70; D. Hanahan and R. A. Weinberg, "Hallmarks of Cancer: the Next Generation," *Cell* 144 (2011): 646–74.

29 *fit into the five categories of cellular cheating in the foundations of multicellular cooperation*   C. A. Aktipis, et al., "Cancer across the Tree of Life: Cooperation and Cheating in Multicellularity." *Philosophical Transasctions of the Royal Society B: Biological Sciences* 370, (2015).

31 *Repeated interactions change the payoffs for cooperation and cheating—often making cooperation a better option overall*   Robert Axelrod and W. D. Hamilton, "The Evolution of Cooperation," *Science* 211, no. 4489 (1981): 1390–96; Robert L. Trivers, "The Evolution of Reciprocal Altruism," *Quarterly Review of Biology* 46, no. 1 (March 1971): 35–57.

31 *When individuals can leave uncooperative partners and groups—or engage in any sort of partner choice*   Ronald Noë and Peter Hammerstein, "Biological Markets: Supply and Demand Determine the Effect of Partner Choice in Cooperation, Mutualism and Mating," *Behavioral Ecology and Sociobiology* 35, no. 1 (1994): 1–11.

31 *As a strategy, cooperation can be better than cheating because groups of cooperators are more stable and longer-lasting*   C. A. Aktipis, "Know When to Walk Away: Contingent Movement and the Evolution of Cooperation," *Journal of Theoretical Biology* 231, no. 2 (2004): 249–60; C. A. Aktipis, "Is Cooperation Viable in Mobile Organisms? Simple Walk Away Rule Favors the Evolution of Cooperation in Groups," *Evolution and Human Behavior* 32, no. 4 (2011): 263–76.

33 *Early multicellular groupings of cells had many other advantages over solitary cells (e.g., the ability to avoid predation and manage risks by sharing and storing*

resources)   John Tyler Bonner, "The Origins of Multicellularity," *Integrative Biology: Issues, News, and Reviews* 1, no. 1 (1998): 27–36.

35   *My colleagues and I have called these the "foundations of multicellular cooperation" in previous publications*   Aktipis, "Cancer across the Tree of Life," 370; A. Aktipis, "Principles of Cooperation across Systems: From Human Sharing to Multicellularity and Cancer," *Evolutionary Applications* 9, no. 1 (2015): 17–36.

35   *In multicellular organisms that are more than a few millimeters across, oxygen and other nutrients can't reach cells on the inside simply through diffusion; some sort of active transport of resources is required*   Andrew H. Knoll and David Hewitt, "Phylogenetic, Functional and Geological Perspectives on Complex Multicellularity," in *The Major Transitions in Evolution Revisited*, ed. Brett Calcott and Kim Sterelny (Cambridge, MA: MIT Press, 2011), 251–70.

37   *a fusion gene called BCR-ABL, in which the promoter (the part of a gene that tells the gene to turn on) from the BCR gene joins with the proliferative signal from another gene, the ABL gene that is responsible for cell proliferation in the immune system*   National Cancer Institute, "NCI Dictionary of Cancer Terms," accessed February 2, 2011. https://www.cancer.gov/publications /dictionaries/cancer-terms.

39   *cooperators can win in the population as a whole, even though they have a disadvantage compared to cheaters within any given group*   Elliott Sober and David Sloan Wilson, *Unto Others: The Evolution and Psychology of Unselfish Behavior* (Cambridge, MA: Harvard University Press, 1998).

41   *there is no debate about the importance of multilevel selection for understanding how cancer cells evolve and how multicellular bodies evolve to suppress and control cancer itself*   Christopher Lean and Anya Plutynski, "The Evolution of Failure: Explaining Cancer as an Evolutionary Process," *Biology and Philosophy* 31, no. 1 (January 2016): 39–57; C. A. Aktipis and R. M. Nesse, "Evolutionary Foundations for Cancer Biology," *Evolutionary Applications* 6, no. 1 (2013): 144–59.

Elephants, for example, have multiple copies of *TP53*, and this is likely one of the reasons that they are particularly resistant to cancer   Lisa M. Abegglen et al., "Potential Mechanisms for Cancer Resistance in Elephants and Comparative Cellular Response to DNA Damage in Humans," *Journal of the American Medical Association* 314, no. 17 (November 2015): 1850–60.

45   *tumor antigen proteins can also be created when the cell cycle is disrupted, when adhesion to neighboring cells is disrupted, and during cellular stress responses*   Olivera J. Finn, "Human Tumor Antigens Yesterday, Today, and Tomorrow," *Cancer Immunology Research* 5, no. 5 (May 2017): 347–54.

45   *The immune system collects information about cellular behavior across all the tissues and organ systems*   Ioana Marin and Jonathan Kipnis, "Learning and Memory . . . and the Immune System," *Learning and Memory* 20, no. 10 (September 2013): 601–6.

46   *the smoke detector principle*   Randolph M. Nesse, "Natural Selection and the Regulation of Defenses: A Signal Detection Analysis of the Smoke Detector Principle," *Evolution and Human Behavior* 26, no. 1 (2005): 88–105.

47 *making it easier to detect the fire (reducing misses), and decrease the number of nuisance alarms (reducing false alarms)* G. Pfister, "Multisensor/Multicriteria Fire Detection: A New Trend Rapidly Becomes State of the Art," *Fire Technology* 33, no. 2 (May 1997): 115–39.

47 *apoptosis may contribute to premature aging (and lead to potential tradeoffs between cancer risk and aging)* Svetlana V. Ukraintseva et al., "Trade-Offs between Cancer and Other Diseases: Do They Exist and Influence Longevity?," *Rejuvenation Research* 13, no. 4 (August 2010): 387–96.

Chapter 4. Cancer from Womb to Tomb

53 *still hotbeds for cancer mutations and precancerous growths* M. Greaves, "Does Everyone Develop Covert Cancer?," *Nature Reviews Cancer* 14, no. 4 (2014): 209–10.

53 *seemingly healthy cells had two to six mutations per million bases, similar to the mutational load found in many cancers* Inigo Martincorena et al., "Tumor Evolution: High Burden and Pervasive Positive Selection of Somatic Mutations in Normal Human Skin," *Science* 348, no. 6237 (May 2015): 880–86.

54 *approximately 0.24 percent of sun-exposed cells were acquiring TP53 mutations each year* Patrik L. Ståhl et al., "Sun-Induced Nonsynonymous p53 Mutations Are Extensively Accumulated and Tolerated in Normal Appearing Human Skin," *Journal of Investigative Dermatology* 131, no. 2 (February 2011): 504–8.

57 *Almost half of conceptions are estimated to fail, 80 percent of which fail before a pregnancy is detectable by standard clinical measures* Kathy Hardy and Philip John Hardy, "1st Trimester Miscarriage: Four Decades of Study," *Translational Pediatrics* 4, no. 2 (April 2015): 189–200.

61 *Humans have diverse mating and marriage patterns, including simultaneously having multiple mates (in the cases of polygyny and polyandry), serially having multiple mates (including serial monogamy, a common pattern for modern Western humans), and sometimes lifelong monogamy as well* Melvin Ember, Carol R. Ember, and Bobbi S. Low, "Comparing Explanations of Polygyny," *Cross-Cultural Research* 41, no. 4 (November 2007): 428–40; Frank W. Marlowe, "The Mating System of Foragers in the Standard Cross-Cultural Sample," *Cross-Cultural Research* 37, no. 3 (August 2003): 282–306; Robert J. Quinlan and Marsha B. Quinlan, "Evolutionary Ecology of Human Pair-Bonds: Cross-Cultural Tests of Alternative Hypotheses," *Cross-Cultural Research* 41, no. 2 (May 2007): 149–69.

62 *David Haig proposed what has come to be known as the "milkshake model," which asks us to imagine a mother who has bought a milkshake for her children to share* David Haig, "Genomic Imprinting and the Theory of Parent-Offspring Conflict," *Seminars in Developmental Biology* 3 (1992): 153–60.

64 *placentas of later-borns tend to be bigger than the placentas of earlier-borns* Thomas McKeown and R. G. Record, "The Influence of Placental Size on Foetal Growth according to Sex and Order of Birth," *Journal of Endocrinology* 10, no. 1 (November 1953): 73–81.

65   *on the other hand, both copies expressing "paternal" growth-promoting gene products leads to a huge placenta*   Wolf Reik et al., "Regulation of Supply and Demand for Maternal Nutrients in Mammals by Imprinted Genes," *Journal of Physiology* 547, pt. 1 (February 2003): 35–44.

65   *Researchers have found that paternally expressed genes contribute to the production of more growth factors and greater placental invasiveness, whereas maternally expressed genes do the opposite*   P. M. Coan, G. J. Burton, and A. C. Ferguson-Smith, "Imprinted Genes in the Placenta—A Review," *Placenta* 26, suppl. A (2005): S10–S20.

65   *gene expression in the placenta (but not the fetus) was dominated by paternally expressed genes*   Xu Wang et al., "Paternally Expressed Genes Predominate in the Placenta," *Proceedings of the National Academy of Sciences of the United States of America* 110, no. 26 (June 2013): 10705–10.

65   *but also for susceptibility to cancer later in life*   David Haig, "Maternal-Fetal Conflict, Genomic Imprinting and Mammalian Vulnerabilities to Cancer," *Philosophical Transactions of the Royal Society of London, Series B: Biological Sciences* 370, no. 1673 (July 2015), https://doi.org/10.1098/rstb.2014.0178.

65   *placental genes driving growth and invasion that should be silent later in life are re-expressed in cancer*   M. Monk and C. Holding, "Human Embryonic Genes Re-Expressed in Cancer Cells," *Oncogene* 20, no. 56 (December 2001): 8085–91.

66   *paternal evolutionary interests can favor cellular phenotypes that are more cancer-like: more proliferative, more invasive, and better able to extract resources from the host*   K. Summers, J. da Silva, and M. A. Farwell, "Intragenomic Conflict and Cancer," *Medical Hypotheses* 59, no. 2 (August 2002): 170–79.

66   *maternally expressed genes produce antibodies that can bind to and inactivate growth factors that are produced by paternally expressed genes*   Haig, "Maternal-Fetal Conflict," 370.

67   *Beckwith-Wiedemann, which is associated with rapid growth in the womb, large size as a child, and an increased risk of cancer*   Haig, "Maternal-Fetal Conflict," 370.

68   *It can take decades for cells with cancerous mutations to proliferate and grow into cancers*   Georg E. Luebeck et al., "Implications of Epigenetic Drift in Colorectal Neoplasia," *Cancer Research* 79, no. 3 (February 2019): 495–504.

68   *mutations that occur early in development can have reverberating effects over the remaining lifespan*   R. Meza, E. G. Luebeck, and S. H. Moolgavkar, "Gestational Mutations and Carcinogenesis," *Mathematical Biosciences* 197, no. 2 (2005): 188–210; S. A. Frank and M. A. Nowak, "Cell Biology: Developmental Predisposition to Cancer," *Nature* 422, no. 6931 (2003): 494.

70   *Women who have their first pregnancy earlier in life are likely to spend less time with undifferentiated stem cells in their breasts*   Benjamin Tiede and Yibin Kang, "From Milk to Malignancy: The Role of Mammary Stem Cells in Development, Pregnancy and Breast Cancer," *Cell Research* 21, no. 2 (February 2011): 245–57.

70   *women who get pregnant earlier in life have a substantially lower risk of hormone-positive breast cancer*   C. Athena Aktipis et al., "Modern Reproductive Patterns Associated with Estrogen Receptor Positive but

Not Negative Breast Cancer Susceptibility," *Evolution, Medicine, and Public Health* 2015, no. 1 (2015): 52–74, https://dx.doi.org/10.1093/emph/eou028; Fabienne Meier-Abt, Mohamed Bentires-Alj, and Christoph Rochlitz, "Breast Cancer Prevention: Lessons to Be Learned from Mechanisms of Early Pregnancy-Mediated Breast Cancer Protection," *Cancer Research* 75, no. 5 (March 2015): 803–7.

70 *cancer cells can evolve to get around this restriction, extending their replicative lives beyond what is optimal for the body* Manuel Collado, Maria A. Blasco, and Manuel Serrano, "Cellular Senescence in Cancer and Aging," *Cell* 130, no. 2 (July 2007): 223–33.

71 *they are also key players in the connections between aging and cancer* Judith Campisi, "Cancer and Ageing: Rival Demons?," *Nature Reviews Cancer* 3, no. 5 (May 2003): 339–49.

71 *the mice that overproduce telomerase exhibit a higher risk for cancer, but if they don't die from cancer, they live longer* Collado, Blasco, and Serrano, "Cellular Senescence in Cancer and Aging," 223–33.

71 *Mice that are deficient in producing telomerase or otherwise have shortened telomeres age more quickly, but also have a lower risk of developing cancer* Collado, Blasco, and Serrano, "Cellular Senescence in Cancer and Aging," 223–33.

71 *when the telomeres of cancer-prone mice are shortened, the risk of cancer goes down* Collado, Blasco, and Serrano, "Cellular Senescence in Cancer and Aging," 223–33.

71 *Killing healthy cells takes those cells out of the population and eventually depletes the renewal capacity of tissues* Campisi, "Cancer and Ageing," 2–13.

71 *mice had lower cancer rates yet did not age more quickly* Campisi, "Cancer and Ageing," 2–13.

72 *cancer is sometimes referred to as "the wound that does not heal"* Harold F. Dvorak, "Tumors: Wounds That Do Not Heal," *New England Journal of Medicine* 315, no. 26 (December 1986): 1650–59.

74 *Leukemias affect a surprisingly large proportion of children before the age of 15* Mel Greaves, "A Causal Mechanism for Childhood Acute Lymphoblastic Leukaemia," *Nature Reviews Cancer* 18, no. 8 (August 2018): 471–84.

74 *the majority of cases of leukemia are diagnosed in adults over the age of 65* "Leukemia—Cancer Stat Facts," Surveillance, Epidemiology, and End Results Program, National Cancer Institute, accessed June 20, 2019, https://seer.cancer.gov/statfacts/html/leuks.html.

74 *The abnormal translocations were already present in the blood of the newborns later diagnosed with leukemia* K. B. Gale et al., "Backtracking Leukemia to Birth: Identification of Clonotypic Gene Fusion Sequences in Neonatal Blood Spots," *Proceedings of the National Academy of Sciences of the United States of America* 94, no. 25 (December 1997): 13950–54.

74 *Approximately 1 percent of newborns have preleukemic clones with these translocations, but only a tiny fraction of them go on to develop clinical ALL* Greaves, "A Causal Mechanism for Childhood Acute Lymphoblastic Leukaemia."

75 *they had fewer exposures to infections in general during early development (compared to those children who were exposed to other children from an early age)* Greaves, "A Causal Mechanism for Childhood Acute Lymphoblastic Leukaemia."

76 *both produce proteins that are responsible for DNA repair, and also play a role in the formation of oocytes (cells in the ovary) and embryonic development* Tuya Pal et al., "Fertility in Women with BRCA Mutations: A Case-Control Study," *Fertility and Sterility* 93, no. 6 (April 2010): 1805–8.

76 BRCA *mutations are associated with many other cancers as well* "BRCA Mutations: Cancer Risk and Genetic Testing," *National Cancer Institute*, February 5, 2018, https://www.cancer.gov/about-cancer/causes-prevention /genetics/brca-fact-sheet.

76 *some* BRCA *mutations are not associated with elevated risk of cancer* Melissa S. Cline et al., "BRCA Challenge: BRCA Exchange as a Global Resource for Variants in BRCA1 and BRCA2," *PLoS Genetics* 14, no. 12 (December 2018): e1007752.

76 *many of these women never receive genetic counseling that might help them interpret their results and better understand their risk* Allison W. Kurian et al., "Gaps in Incorporating Germline Genetic Testing into Treatment Decision-Making for Early-Stage Breast Cancer," *Journal of Clinical Oncology* 35, no. 20 (July 2017): 2232–39.

77 *there are many possible mutations in the* BRCA1 *and* BRCA2 *genes, and some of these mutations contribute to our risk of cancer* Cline et al., "BRCA Challenge."

77 BRCA *mutations confer a 65–80 percent risk of breast cancer in women who harbor these mutations, compared to a 12–13 percent risk in the general female population* Hagit Daum, Tamar Peretz, and Neri Laufer, "BRCA Mutations and Reproduction," *Fertility and Sterility* 109, no. 1 (January 2018): 33–38.

77 *About 25 percent of* BRCA1 *mutation carriers are diagnosed with breast cancer before the age of forty, and 72 percent are diagnosed by the age of eighty* Karoline B. Kuchenbaecker et al., "Risks of Breast, Ovarian, and Contralateral Breast Cancer for BRCA1 and BRCA2 Mutation Carriers," *Journal of the American Medical Association* 317, no. 23 (June 2017): 2402–16.

77 BRCA *mutations are not limited to women; men with* BRCA *mutations have increased risk of breast and prostate cancer* Daum, Peretz, and Laufer, "BRCA Mutations and Reproduction," 33–38.

78 *They found that women with* BRCA *mutations had female ancestors that had more offspring—with an average of 1.9 more offspring compared to the ancestors of women without these* BRCA *mutations (for women born before 1930 the controls had on average 4.19 offspring, whereas carriers had an average of 6.22 offspring)* K. R. Smith et al., "Effects of BRCA1 and BRCA2 Mutations on Female Fertility," *Proceedings of the Royal Society of London, Series B* 279, no. 1732 (2011): 1389–95, https://doi.org/10.1098/rspb.2011.1697.

78 *Women with* BRCA *mutations in this sample had more children (1.8 more on average compared to controls), were less likely to have no children, and had lower rates of miscarriage* Fabrice Kwiatkowski et al., "BRCA Mutations

Increase Fertility in Families at Hereditary Breast/Ovarian Cancer Risk," *PloS One* 10, no. 6 (June 2015): e0127363.

78 *a study of women from the United States and Canada did not find a significant relationship between fertility and* BRCA *mutations* Pal et al., "Fertility in Women with BRCA Mutations."

78 *found no evidence of increased fertility in women with* BRCA *mutations, though the researchers did find that women with* BRCA *mutations had more female offspring (almost 60 percent females) than women without these mutations (who had just over 50 percent female offspring)* Roxana Moslehi et al., "Impact of BRCA Mutations on Female Fertility and Offspring Sex Ratio," *American Journal of Human Biology* 22, no. 2 (March 2010): 201–5.

79 *Many of these mutations are associated with increased risk of breast and ovarian cancer—though the risk of cancer varies for the particular mutation and for the particular population* Brad Keoun, "Ashkenazim Not Alone: Other Ethnic Groups Have Breast Cancer Gene Mutations, Too," *Journal of the National Cancer Institute* 89, no. 1 (January 1997): 8–9.

81 *it helps suppress metastasis in breast cancer and melanoma* Aktipis et al., "Modern Reproductive Patterns Associated with Estrogen Receptor Positive but Not Negative Breast Cancer Susceptibility."

81 *high levels of testosterone are also correlated with higher risk of prostate cancer in the long term* L. C. Alvarado, "Do Evolutionary Life-History Trade-Offs Influence Prostate Cancer Risk? A Review of Population Variation in Testosterone Levels and Prostate Cancer Disparities," *Evolutionary Applications* 6, no. 1 (2013): 117–33.

82 *Cancer defense only paid off when extrinsic mortality (the likelihood of dying from random causes) was low and competitiveness made little difference to reproductive success (when it was not a "winner-take-all" mating system)* A. M. Boddy et al., "Cancer Susceptibility and Reproductive Trade-Offs: A Model of the Evolution of Cancer Defences," *Philosophical Transactions of the Royal Society of London, Series B: Biological Sciences* 370, no. 1673 (2015), https://doi.org/10.1098/rstb.2014.0220.

82 *Traits like faster cell proliferation, sloppier DNA repair, and more permissive conditions for conception and/or implantation of an embryo can provide an organism-level advantage in terms of reproductive competitiveness, but may come at a cost in terms of cancer susceptibility* Boddy et al., "Cancer Susceptibility and Reproductive Trade-Offs," 370.

84 *We live with precancerous growth for decades, usually without any problems* Greaves, "Does Everyone Develop Covert Cancer?"

84 *our hunter-gatherer ancestors likely invested in their offspring and even grand-offspring for decades after birth* K. Hawkes et al., "Grandmothering, Menopause, and the Evolution of Human Life Histories," *Proceedings of the National Academy of Sciences of the United States of America* 95, no. 3 (February 1998): 1336–39.

85 *selection for cancer suppression can remain high enough after reproduction to favor cancer suppression mechanisms in old age* J. S. Brown and C. A. Aktipis,

"Inclusive Fitness Effects Can Select for Cancer Suppression into Old Age," *Philosophical Transactions of the Royal Society of London, Series B: Biological Sciences* 370, no. 1673 (2015), https://doi.org/10.1098/rstb.2015.0160.

85   *Data from modern hunter-gatherers show that humans in conditions similar to those of our ancestors often live past seventy years old*   M. Gurven and H. Kaplan, "Longevity among Hunter-Gatherers: A Cross-Cultural Examination," *Population and Development Review* 33, no. 2 (2007): 321–65, https://onlinelibrary.wiley.com/doi/abs/10.1111/j.1728-4457.2007.00171.x.

85   *we are less likely to die of other causes early in life—like accidents and infections*   Office for National Statistics, "Causes of Death over 100 Years," September 18, 2017, https://www.ons.gov.uk/peoplepopulationandcommunity/birthsdeath sandmarriages/deaths/articles/causesofdeathover100years/2017-09-18.

85   *In addition to increased risk from more calories and more sedentary behavior*   Véronique Bouvard et al., "Carcinogenicity of Consumption of Red and Processed Meat," *Lancet Oncology* 16, no. 16 (December 2015): 1599–1600.

85   *thanks to modern conveniences, our life is associated with other exposures like chemical carcinogens*   "Carcinogens Listed in the Eleventh Report," in *The Report on Carcinogens*, 11th ed. (Durham, NC: National Toxicology Program, U.S. Department of Health and Human Services, 2011), https://web.archive .org/web/20090507123840if_/http://ntp.niehs.nih.gov/ntp/roc/eleventh /known.pdf.

85   *higher levels of reproductive hormones (because of better nutrition*   Alvarado, "Do Evolutionary Life-History Trade-Offs Influence Prostate Cancer Risk?"

85   *and, for women, more frequent ovulation*   F. Clavel-Chapelon and E3N Group, "Cumulative Number of Menstrual Cycles and Breast Cancer Risk: Results from the E3N Cohort Study of French Women," *Cancer Causes and Control* 13, no. 9 (November 2002): 831–38.

85   *and greater disruption to our sleep*   S. Davis, D. K. Mirick, and R. G. Stevens, "Night Shift Work, Light at Night, and Risk of Breast Cancer," *Journal of the National Cancer Institute* 93, no. 20 (October 2001): 1557–62.

Chapter 5. Cancer across the Tree of Life

87   *Dogs with chronic myeloid leukemias have even been found to have the* BCR/ABL *translocation, the same translocation that I discussed in chapter 3, which is typical of chronic myeloid leukemias in humans*   Joshua D. Schiffman and Matthew Breen, "Comparative Oncology: What Dogs and Other Species Can Teach Us about Humans with Cancer," *Philosophical Transactions of the Royal Society of London, Series B: Biological Sciences* 370, no. 1673 (July 2015), https://doi.org/10.1098/rstb.2014.0231.

88   *In both dogs and humans, the risk of cancer is associated with larger size*   J. M. Fleming, K. E. Creevy, and D. E. L. Promislow, "Mortality in North American Dogs from 1984 to 2004: An Investigation into Age-, Size-, and Breed-Related Causes of Death," *Journal of Veterinary Internal Medicine* 25, no. 2 (March 2011): 187–98; Jane Green et al., "Height and Cancer Incidence in the Million Women Study: Prospective Cohort, and Meta-Analysis of

Prospective Studies of Height and Total Cancer Risk," *Lancet Oncology* 12, no. 8 (August 2011): 785–94; Sara Wirén et al., "Pooled Cohort Study on Height and Risk of Cancer and Cancer Death," *Cancer Causes and Control* 25, no. 2 (February 2014): 151–59.

88   *their rates of cancer are much lower than ours*   Lisa M. Abegglen et al., "Potential Mechanisms for Cancer Resistance in Elephants and Comparative Cellular Response to DNA Damage in Humans," *Journal of the American Medical Association* 314, no. 17 (November 2015): 1850–60.

88   *Peto's Paradox*   R. Peto et al., "Cancer and Ageing in Mice and Men," *British Journal of Cancer* 32, no. 4 (October 1975): 411–26; Richard Peto, "Epidemiology, Multistage Models, and Short-Term Mutagenicity Tests," *International Journal of Epidemiology* 45, no. 3 (1977): 621–37.

90   *fasciation as result of damage on a cypress tree, Chamaecyparis obtuse*   Anton Baudoin, Virginia Polytechnic Institute and State University; Bugwood.org is licensed under CC BY 3.0.

90   *a crested Casuarina glauca, which lacks the typical branching structure and instead features a large fan of tissue with dysregulated differentiation*   Tyler ser Noche, File:Starr-180421-0291-Casuarina glauca-with fasciated branch-Honolua Lipoa Point-Maui (41651326770).jpg is licensed under CC BY 3.0.

90   *a mule's ear flower, Wyenthia helianthoides, with a normal flower shown on the left and fasciated form on the right*   Perduejn, Mules Ear Fasciated is licensed under CC BY 3.0.

90   *a "double flower"* Anemone coronaria   Thomas Bresson, 2014-03-09 14-30-31 fleur-18f is licensed under CC BY 3.0.

92   *All branches of multicellular life are susceptible to cancer. In our review of cancer across the tree of life, we found reports of cancer and cancer-like phenomena (dysregulated differentiation and overproliferation) across every branch of multicellularity*   Figure reprinted with permission, Aktipis 2015, licensed by CC BY 4.0); C. A. Aktipis et al. "Cancer across the Tree of Life: Cooperation and Cheating in Multicellularity." *Philosophical Transasctions of the Royal Society B: Biological Sciences* 370 (2015).

93   *an invading front of cells breaking through existing tissues*   G. N. Agrios, *Plant Pathology* (Boston: Elsevier Academic Press, 2005), 922; Philip R. White and Armin C. Braun, "A Cancerous Neoplasm of Plants: Autonomous Bacteria-Free Crown-Gall Tissue," *Cancer Research* 2, no. 9 (1942): 597–617.

95   *cells in the neighboring areas can create a ring of actomyosin (what muscles are made of) that literally squeezes out the problematic cells*   George T. Eisenhoffer et al., "Crowding Induces Live Cell Extrusion to Maintain Homeostatic Cell Numbers in Epithelia," *Nature* 484, no. 7395 (April 2012): 546–49.

95   *normal cells can produce filamin and vimentin that create long arm-like protrusions that expel mutated cells*   Mihoko Kajita et al., "Filamin Acts as a Key Regulator in Epithelial Defence against Transformed Cells," *Nature Communications* 5 (July 2014): 4428.

95   *This mechanism only works, however, if the cells around the mutated cells are normal*   Kajita et al., "Filamin Acts as a Key Regulator in Epithelial Defence against Transformed Cells," 4428.

96 *larger breeds of dogs (heavier than about 20 kilograms or 44 pounds) have a greater risk of cancer than dogs from smaller breeds* Fleming, Creevy, and Promislow, "Mortality in North American Dogs from 1984 to 2004."

96 *taller humans have a greater risk of cancer than shorter humans—with about a 10 percent increased risk of cancer for every 10 centimeters (about 3.9 inches) of height* Green et al., "Height and Cancer Incidence in the Million Women Study"; Wirén et al., "Pooled Cohort Study on Height and Risk of Cancer and Cancer Death."

96 *Yet this same pattern of greater cancer risk with larger size does not hold if we look across species* Leonard Nunney et al., "Peto's Paradox and the Promise of Comparative Oncology," *Philosophical Transactions of the Royal Society of London, Series B: Biological Sciences* 370, no. 1673 (July 2015), https://doi.org/10.1098/rstb.2014.0177.

97 *on a cell-to-cell comparison—human cells must be more cancer resistant than mouse cells, otherwise we would succumb to cancer at an early age* Peto, "Epidemiology, Multistage Models, and Short-Term Mutagenicity Tests."

97 *Species with longer lifespans and larger sizes do not have higher cancer incidence than species that are smaller and shorter-lived* Abegglen et al., "Potential Mechanisms for Cancer Resistance in Elephants and Comparative Cellular Response to DNA Damage in Humans"; A. F. Caulin and C. C. Maley. "Peto's Paradox: Evolution's Prescription for Cancer Prevention." *Trends in Ecology and Evolution* 26, no. 4 (February 2011): 175–82.

99 *In addition to producing more eggs, they also have high rates of ovarian cancer, likely because they have been selected to have more permissive cell proliferation in and around their ovaries* Boddy et al., "Cancer Susceptibility and Reproductive Trade-Offs"; P. A. Johnson and J. R. Giles, "The Hen as a Model of Ovarian Cancer," *Nature Reviews Cancer* 13, no. 6 (2013): 432–36.

99 *Then, during the spring and summer, the antlers grow back rapidly in preparation for the breeding season in the fall* Robert A. Pierce II, Jason Sumners, and Emily Flinn, "Antler Development in White-Tailed Deer: Implications for Management," University of Missouri Extension, January 2012, https://extension2.missouri.edu/g9486.

100 *Even normal antlers without antleromas have gene expression patterns more indicative of bone cancer (osteosarcoma) than normal bone* Yu Wang et al., "Genetic Basis of Ruminant Headgear and Rapid Antler Regeneration," *Science* 364, no. 6446 (June 2019), https://doi.org/10.1126/science.aav6335.

100 *genetic sequencing has revealed that genes associated with cancer (proto oncogenes) have been under positive selection in the ancestors of cervids (deer)* Wang et al., "Genetic Basis of Ruminant Headgear and Rapid Antler Regeneration."

100 *These antlers are an example of a sexually selected trait (because females are more likely to mate with males with larger antlers) that increases susceptibility to cancer* Boddy et al., "Cancer Susceptibility and Reproductive Trade-Offs."

101 *The same gene that is responsible for large size also makes these fish more susceptible to skin cancer* André A. Fernandez and Paul R. Bowser,

"Selection for a Dominant Oncogene and Large Male Size as a Risk Factor for Melanoma in the Xiphophorus Animal Model," *Molecular Ecology* 19, no. 15 (August 2010): 3114–23.

101   *elephants get an extra dose of all this cancer suppression functionality; they are especially sensitive to DNA damage, so their cells self-destruct more readily if damage occurs*   Abegglen et al., "Potential Mechanisms for Cancer Resistance in Elephants and Comparative Cellular Response to DNA Damage in Humans."

103   *The team combined Maley's analyses showing that elephants have 40 copies of TP53 with the data from Schiffman's lab showing that elephant cells readily self-destruct in response to radiation*   Abegglen et al., "Potential Mechanisms for Cancer Resistance in Elephants and Comparative Cellular Response to DNA Damage in Humans."

103   *over the course of organismal evolutionary time the number of copies of TP53 increased when body size increased*   Michael Sulak et al., "TP53 Copy Number Expansion Is Associated with the Evolution of Increased Body Size and an Enhanced DNA Damage Response in Elephants," *eLife* 5 (September 2016), https://doi.org/10.7554/eLife.11994.

103   *humpback whales have duplications of apoptosis genes; and have had positive selection on genes responsible for cell cycle control, cell signaling, and cell proliferation compared to their smaller cetacean cousins (including the sperm whale, bottlenose dolphin, and orca)*   Marc Tollis et al., "Return to the Sea, Get Huge, Beat Cancer: An Analysis of Cetacean Genomes Including an Assembly for the Humpback Whale (Megaptera Novaeangliae)," *Molecular Biology and Evolution* 36, no. 8 (August 2019): 1746–63.

104   *those genes that live between the "unicellular" freedom-promoting genes and the "multicellular" control-promoting genes*   Anna S. Trigos et al., "Altered Interactions between Unicellular and Multicellular Genes Drive Hallmarks of Transformation in a Diverse Range of Solid Tumors," *Proceedings of the National Academy of Sciences of the United States of America* 114, no. 24 (June 2017): 6406–11.

104   *The gatekeeper genes that exist at the interface between "unicellular" and "multicellular" genes are the most evolutionarily recent*   Trigos et al., "Altered Interactions between Unicellular and Multicellular Genes Drive Hallmarks of Transformation in a Diverse Range of Solid Tumors."

105   *some cells specialized in trying to invade and take advantage of these cooperative cellular societies rather than creating and maintaining a cellular society of their own*   Richard K. Grosberg and Richard R. Strathmann, "The Evolution of Multicellularity: A Minor Major Transition?," *Annual Review of Ecology, Evolution, and Systematics* 38, no. 1 (December 2007): 621–54.

105   *for multicellular life to be viable, early organisms had to evolve ways of keeping these invaders out*   Leo W. Buss, *The Evolution of Individuality* (Princeton, NJ: Princeton University Press, 1987); L. W. Buss, "Somatic Cell Parasitism and the Evolution of Somatic Tissue Compatibility," *Proceedings of the National Academy of Sciences of the United States of America* 79, no. 17 (September 1982): 5337–41.

105 *Our immune system also includes the skin, which helps protect us from external threats* Ehrhardt Proksch, Johanna M. Brandner, and Jens-Michael Jensen, "The Skin: An Indispensable Barrier," *Experimental Dermatology* 17, no. 12 (December 2008): 1063–72.

106 *CTVT was so successful that it may have even led to the extinction of the first dogs in North America* Angela Perri et al., "New Evidence of the Earliest Domestic Dogs in the Americas," *bioRxiv*, June 27, 2018, https://doi.org/10.1101/343574.

106 *When the dogs try to separate, the genital area can be injured, breaching the first line of defense of the immune system—the skin* E. P. Murchison, "Clonally Transmissible Cancers in Dogs and Tasmanian Devils," *Oncogene* 27, suppl. 2 (December 2008): S19–S30.

107 *CTVT is also the only known unicellular species of dog* Clare A. Rebbeck et al., "Origins and Evolution of a Transmissible Cancer," *Evolution: International Journal of Organic Evolution* 63, no. 9 (September 2009): 2340–49.

107 *much like other unicellular infectious agents, surviving long after their hosts have died and continuing to successfully transmit to other members of the host population* Murchison, "Clonally Transmissible Cancers in Dogs and Tasmanian Devils."

107 *a strange facial tumor in Tasmanian devils in the northeast corner of Tasmania, an island off of Australia's southeast coast.* Murchison, "Clonally Transmissible Cancers in Dogs and Tasmanian Devils."

108 *the host has countless opportunities to transmit the facial tumor to other devils* Murchison, "Clonally Transmissible Cancers in Dogs and Tasmanian Devils."

108 *The second one discovered, DFTD2, has a Y-chromosome, indicating that it originated from a male* Ruth J. Pye et al., "A Second Transmissible Cancer in Tasmanian Devils," *Proceedings of the National Academy of Sciences of the United States of America* 113, no. 2 (January 2016): 374–79.

109 *Contagious cancers may not be as rare and bizarre as we've thought* Pye et al., "A Second Transmissible Cancer in Tasmanian Devils."

109 *Both dogs and devils have gone through genetic bottlenecks, periods in history when the genetic diversity of the population went down* Murchison, "Clonally Transmissible Cancers in Dogs and Tasmanian Devils."

109 *largely because of campaigns to eliminate them on the part of European settlers who arrived in Tasmania in the nineteenth century* Hannah V. Siddle and Jim Kaufman, "A Tale of Two Tumours: Comparison of the Immune Escape Strategies of Contagious Cancers," *Molecular Immunology* 55, no. 2 (September 2013): 190–93.

110 *allowing the immune system to tell self from nonself* Siddle and Kaufman, "A Tale of Two Tumours."

110 *regression is associated with an increase in MHC expression and the presence of immune cells at the site of the tumor* Siddle and Kaufman, "A Tale of Two Tumours."

111   *activation of blood cells called hemocytes that play a role in detecting and responding to potential infection*   Bassem Allam and David Raftos, "Immune Responses to Infectious Diseases in Bivalves," *Journal of Invertebrate Pathology* 131 (October 2015): 121–36.

111   *transmissible cancer may be responsible for bivalve leukemia-like cancers more generally*   Michael J. Metzger and Stephen P. Goff, "A Sixth Modality of Infectious Disease: Contagious Cancer from Devils to Clams and Beyond," *PLoS Pathogens* 12, no. 10 (October 2016): e1005904.

112   *transmissible cancers have been a selective pressure on organisms since the beginnings of multicellular life*   Metzger and Goff, "A Sixth Modality of Infectious Disease."

112   *Genetic analysis showed that they were actually tapeworm cells, growing as a cancer in the tissue of the patient*   Atis Muehlenbachs et al., "Malignant Transformation of Hymenolepis Nana in a Human Host," *New England Journal of Medicine* 373, no. 19 (November 2015): 1845–52.

113   *Three of the four reports of this tapeworm cancer noted that the patient was HIV-positive*   Muehlenbachs et al., "Malignant Transformation of Hymenolepis Nana in a Human Host"; Peter D. Olson et al., "Lethal Invasive Cestodiasis in Immunosuppressed Patients," *Journal of Infectious Diseases* 187, no. 12 (June 2003): 1962–66; M. Santamaría-Fríes et al., "Lethal Infection by a Previously Unrecognised Metazoan Parasite," *Lancet* 347, no. 9018 (June 1996): 1797–1801.

113   *the fourth individual had a compromised immune system due to Hodgkin's lymphoma*   Olson et al., "Lethal Invasive Cestodiasis in Immunosuppressed Patients."

113   *Organ transplants save thousands of lives every year*   Health Resources and Services Administration, "Organ Donation Statistics," accessed December 19, 2017, https://www.organdonor.gov/statistics-stories/statistics.html.

114   *He noticed an apparent increase in cancer among transplant recipients*   I. Penn, C. G. Halgrimson, and T. E. Starzl, "De Novo Malignant Tumors in Organ Transplant Recipients," *Transplantation Proceedings* 3, no. 1 (March 1971): 773–78.

114   *and created a registry for transplanted tumors*   Beth Ann Witherow et al., "The Israel Penn International Transplant Tumor Registry," *AMIA Annual Symposium Proceedings* (2003): 1053.

114   *a study of more than one hundred thousand donors found only eighteen cases, making the donor-related tumor rate for transplanted cadaveric organs incredibly small, at 0.017 percent*   H. Myron Kauffman et al., "Transplant Tumor Registry: Donor Related Malignancies," *Transplantation* 74, no. 3 (August 2002): 358–62.

114   *a study of hundreds of donors with CNS tumors found no examples of transmission*   H. Myron Kauffman et al., "Transplant Tumor Registry: Donors with Central Nervous System Tumors," *Transplantation* 73, no. 4 (February 2002): 579–82.

114 *The benefits of getting an organ transplant far outweigh the risk of getting a transmissible cancer from the procedure* E. F. Scanlon et al., "Fatal Homotransplanted Melanoma: A Case Report," *Cancer* 18 (June 1965): 782–89; A. E. Moore, C. P. Rhoads, and C. M. Southam, "Homotransplantation of Human Cell Lines," *Science* 125, no. 3239 (January 1957): 158–60.

114 *screening is constantly being improved and updated to better prevent the inadvertent transplanting of tumors from donors to recipients* Manish J. Gandhi and D. Michael Strong, "Donor Derived Malignancy Following Transplantation: A Review," *Cell and Tissue Banking* 8, no. 4 (April 2007): 267–86.

114 *Genetic testing showed that the tumor originated from the patient that the surgeon had operated on* H. V. Gärtner et al., "Genetic Analysis of a Sarcoma Accidentally Transplanted from a Patient to a Surgeon," *New England Journal of Medicine* 335, no. 20 (November 1996): 1494–96.

115 *a laboratory worker accidentally punctured herself with a needle containing colonic adenocarcinoma cells* E. A. Gugel and M. E. Sanders, "Needle-Stick Transmission of Human Colonic Adenocarcinoma," *New England Journal of Medicine* 315, no. 23 (December 1986): 1487.

115 *cancer biologist Mel Greaves estimates the chance that a pregnant woman with cancer would transmit that cancer to her fetus to be around one in five hundred thousand* Mel Greaves and William Hughes, "Cancer Cell Transmission via the Placenta," *Evolution, Medicine, and Public Health* 2018, no. 1 (April 2018): 106–15.

115 *there are also many reports of leukemia being transmitted between monozygotic twins in the womb* Mel F. Greaves et al., "Leukemia in Twins: Lessons in Natural History," *Blood* 102, no. 7 (October 2003): 2321–33.

116 *one of the cornerstones of our vertebrate immune system may have arisen because of the selective pressures that came from transmissible cancers* Murchison, "Clonally Transmissible Cancers in Dogs and Tasmanian Devils"; Katherine Belov, "The Role of the Major Histocompatibility Complex in the Spread of Contagious Cancers," *Mammalian Genome* 22, no. 1–2 (February 2011): 83–90; Claudio Murgia et al., "Clonal Origin and Evolution of a Transmissible Cancer," *Cell* 126, no. 3 (August 2006): 477–87; Sven Kurbel, Stjepko Plestina, and Damir Vrbanec, "Occurrence of the Acquired Immunity in Early Vertebrates due to Danger of Transmissible Cancers Similar to Canine Venereal Tumors," *Medical Hypotheses* 68, no. 5 (2007): 1185–86.

116 *One of the prevailing theories for the evolution of sex is that it creates genetic diversity that makes offspring less vulnerable to transmission of infections* W. D. Hamilton, R. Axelrod, and R. Tanese, "Sexual Reproduction as an Adaptation to Resist Parasites (a Review)," *Proceedings of the National Academy of Sciences of the United States of America* 87, no. 9 (May 1990): 3566–73.

117 *the theory states that sexual reproduction can increase genetic heterogeneity in the population, decreasing the vulnerability of offspring to transmissible cancer* Frédéric Thomas et al., "Transmissible Cancer and the Evolution of Sex," *PLoS Biology* 17, no. 6 (June 2019): e3000275.

117 *If sexual reproduction did evolve in part because it reduced the risk of contagious cancer, it is certainly ironic that dog infectious cancers are transmitted through*

117 *sexual contact*   Murchison, "Clonally Transmissible Cancers in Dogs and Tasmanian Devils."

117 *transmissible cancer may be the reason for the extinction of the first dogs in North America*   Perri et al., "New Evidence of the Earliest Domestic Dogs in the Americas."

Chapter 6. The Hidden World of Cancer Cells

120 *cancer cells can behave like normal cells if you put them into an environment with normal cells*   M. J. Bissell and W. C. Hines, "Why Don't We Get More Cancer? A Proposed Role of the Microenvironment in Restraining Cancer Progression," *Nature Medicine* 17, no. 3 (2011): 320–29.

120 *The centrality of the tumor microenvironment to keeping cancer at bay is also the main idea of the "tissue organization field theory" of cancer*   C. Sonnenschein and A. M. Soto, *The Society of Cells: Cancer and Control of Cell Proliferation* (New York: Springer, 1999).

121 *Chronic inflammation is one common characteristic of tumor microenvironments*   M. J. Bissell and D. Radisky, "Putting Tumours in Context," *Nature Reviews Cancer* 1, no. 1 (2001): 46–54.

122 *billions of new nucleotides must be synthesized for a cell to reproduce, leading to tremendous demand for nitrogen and phosphorus at tumor sites*   James J. Elser et al., "Biological Stoichiometry in Human Cancer," *PloS One* 2, no. 10 (2007): e1028.

122 *one way to do this is for the cancer cells to send out wound-healing signals to these support cells, which then prompts them to send back growth and survival factors*   Robert A. Gatenby and Robert J. Gillies, "A Microenvironmental Model of Carcinogenesis," *Nature Reviews Cancer* 8, no. 1 (January 2008): 56–61.

123 *they produce factors that shut down proliferation, induce cell death, and block the recruitment of blood vessels to cut off the tumor from resources*   Gavin P. Dunn et al., "Cancer Immunoediting: From Immunosurveillance to Tumor Escape," *Nature Immunology* 3, no. 11 (November 2002): 991–98.

123 *This is very much like how prey animals evolve to evade predators*   Aktipis and Nesse, "Evolutionary Foundations for Cancer Biology."

123 *cancer cells evolve to evade the immune system through strategies like hiding (removing the markers on the outside of the cell that immune cells can identify) and camouflage (expressing genes that give them a more "normal" appearance to immune cells)*   D. Gabrilovich and V. Pisarev, "Tumor Escape from Immune Response: Mechanisms and Targets of Activity," *Current Drug Targets* 4, no. 7 (2003): 525–36; F. Cavallo et al., "2011: The Immune Hallmarks of Cancer," *Cancer Immunology, Immunotherapy* 60, no. 3 (2011): 319–26.

123 *they can create conditions that favor cells that can disperse in order to find new environments to colonize, driving invasion and metastasis*   C. A. Aktipis, C. C. Maley, and J. W. Pepper, "Dispersal Evolution in Neoplasms: The Role of Disregulated Metabolism in the Evolution of Cell Motility," *Cancer Prevention Research* 5, no. 2 (2012): 266–75.

124   *cancer cells evolve to have different life history strategies depending on the environments they are in*   C. A. Aktipis et al., "Life History Trade-Offs in Cancer Evolution," *Nature Reviews Cancer* 13, no. 12 (2013): 883–92.

125   *trade-offs between proliferation and survival become more important*   Aktipis et al., "Life History Trade-Offs in Cancer Evolution."

125   *Approximately half of all the energy of the cell is required to operate these pumps*   H. J. Broxterman et al., "Induction by Verapamil of a Rapid Increase in ATP Consumption in Multidrug-Resistant Tumor Cells," *FASEB Journal* 2, no. 7 (April 1988): 2278–82.

126   *Breaking through this barrier and others often requires cancer cells to cooperate with each other to produce factors (called matrix metalloproteinases) that break up the membrane*   Anna Chapman et al., "Heterogeneous Tumor Subpopulations Cooperate to Drive Invasion," *Cell Reports* 8, no. 3 (August 2014): 688–95.

126   *Getting through the body's membranes and other tissues also requires that cancer cells coordinate their electrical signaling*   Guangping Tai, Michael Tai, and Min Zhao, "Electrically Stimulated Cell Migration and Its Contribution to Wound Healing," *Burns and Trauma* 6 (July 9, 2018): 20; Anna Haeger et al., "Collective Cell Migration: Guidance Principles and Hierarchies," *Trends in Cell Biology* 25, no. 9 (September 2015): 556–66.

126   *remodeling the tissue architecture (for example, producing collagen, which can make tumors feel like nodules in otherwise elastic tissues) and signaling for new blood vessels*   Gatenby and Gillies, "A Microenvironmental Model of Carcinogenesis."

126   *the process of niche construction can involve a twisted form of cellular cooperation between cancer cells and apparently normal cells*   Gatenby and Gillies, "A Microenvironmental Model of Carcinogenesis."

130   *the high resource consumption of cancer cells leads to selection for cells that move*   Aktipis, Maley, and Pepper, "Dispersal Evolution in Neoplasms."

131   *metastases often arise from cancer cells that left a tumor early in progression, in a process called "early dissemination"*   Hedayatollah Hosseini et al., "Early Dissemination Seeds Metastasis in Breast Cancer," *Nature* 540, no. 7634 (December 2016): 552–58.

131   *the act of invasion often requires cancer cells to cooperate with each other to produce matrix metalloproteinases that break up the basement membrane*   Chapman et al., "Heterogeneous Tumor Subpopulations Cooperate to Drive Invasion."

131   *cancer cells can "trick" the vascular system into letting them through the endothelium (the membrane around the blood vessels) by coordinating their electrical signaling*   Tai, Tai, and Zhao. "Electrically Stimulated Cell Migration and Its Contribution to Wound Healing"; Haeger et al., "Collective Cell Migration."

133   *By cooperating, they can get the benefits from essentially dividing the labor of producing all these factors*   Lee Alan Dugatkin, "Animal Cooperation among Unrelated Individuals," *Die Naturwissenschaften* 89, no. 12 (December 2002):

533–41; R. Axelrod, D. E. Axelrod, and K. J. Pienta, "Evolution of Cooperation among Tumor Cells," *Proceedings of the National Academy of Sciences of the United States of America* 103, no. 36 (2006): 13474–79; Marco Archetti, "Cooperation between Cancer Cells," *Evolution, Medicine, and Public Health* 2018, no. 1 (January 2018): 1.

133 *Positive assortment is when cooperators are more likely to interact with each other than with a random individual in the population* J. A. Fletcher and Michael Doebeli, "A Simple and General Explanation for the Evolution of Altruism," *Proceedings of the Royal Society B: Biological Sciences* 276, no. 1654 (2009): 13–19.

133 *and positive assortment can favor the evolution of cooperation, regardless of whether the cooperation is among relatives, repeated interaction partners or even members of different species who provide benefits for each other* J. A. Fletcher and Michael Doebeli, "A Simple and General Explanation for the Evolution of Altruism," *Proceedings of the Royal Society B: Biological Sciences* 276, no. 1654 (2009): 13–19.

133 *Some cancer biologists are skeptical about the importance of kin selection as an explanation for cooperation among cancer cells* Axelrod, Axelrod, and Pienta, "Evolution of Cooperation among Tumor Cells."

134 *But it is also possible for cooperation to evolve among cells that produce different growth factors produced by different genes, through the same process of positive assortment that I discussed earlier* Archetti and Pienta, "Cooperation among Cancer Cells."

134 *Cooperation can evolve as long as cooperators preferentially interact with one another* Fletcher and Doebeli, "A Simple and General Explanation for the Evolution of Altruism."

134 *it isn't necessary to recognize who is and who is not kin for investment in relatives to be effective* C. A. Aktipis and E. Fernandez-Duque, "Parental Investment without Kin Recognition: Simple Conditional Rules for Parent–Offspring Behavior," *Behavioral Ecology and Sociobiology* 65, no. 5 (May 2011): 1079–91.

134 *Cooperation among relatives can evolve via kin selection when there is a kin structure (like offspring staying near parents) in which recipients of benefits are likely to be relatives* W. D. Hamilton, "The Genetical Evolution of Social Behaviour. I," *Journal of Theoretical Biology* 7, no. 1 (July 1964): 1–16; W. D. Hamilton, "The Genetical Evolution of Social Behaviour. II," *Journal of Theoretical Biology* 7, no. 1 (July 1964): 1–16.

135 *somewhere between 75 percent and 99.999 percent of cells in tumors are cancer nonstem cells and cannot propagate the tumor* Kathleen Sprouffske et al., "An Evolutionary Explanation for the Presence of Cancer Nonstem Cells in Neoplasms," *Evolutionary Applications* 6, no. 1 (January 2013): 92–101.

135 *we found that these cells with limited potential to divide were maintained in the population of cells* Kathleen Sprouffske et al., "An Evolutionary Explanation for the Presence of Cancer Nonstem Cells in Neoplasms," *Evolutionary Applications* 6, no. 1 (January 2013): 92–101.

135 *(this phenomenon occurs with many birds who have related helpers at the nest)* Ralph Bergmüller et al., "Integrating Cooperative Breeding into Theoretical Concepts of Cooperation," *Behavioural Processes* 76, no. 2 (2007): 61–72.

136 *Being social makes sense when resources are unpredictable, because sociality provides a buffer against the challenges of the variable environment* Bert Hölldobler and Edward O. Wilson, *The Superorganism: The Beauty, Elegance, and Strangeness of Insect Societies* (New York: W. W. Norton and Company, 2009); Lee Cronk et al., "Managing Risk through Cooperation: Need-Based Transfers and Risk Pooling among the Societies of the Human Generosity Project," in *Global Perspectives on Long-Term Community Resource Management*, ed. L. Lozny and T. McGovern (New York: Springer, 2019), 41–75.

136 *they deeply change the structure of the ecological community around them, often displacing native species* Henrik Moller, "Lessons for Invasion Theory from Social Insects," *Biological Conservation* 78, no. 1 (October 1996): 125–42.

137 *In a mouse model of mammary tumors, circulating tumor cell clusters were twenty-three to fifty times more likely to successfully create metastases than individual cells* Nicola Aceto et al., "Circulating Tumor Cell Clusters Are Oligoclonal Precursors of Breast Cancer Metastasis," *Cell* 158, no. 5 (2014): 1110–22.

137 *certain polyclonal tumors (containing multiple different clones) can have a proliferation advantage, because they contain clones that can collectively colonize and maintain the cancer niche* Andriy Marusyk et al., "Non-Cell-Autonomous Driving of Tumour Growth Supports Sub-Clonal Heterogeneity," *Nature* 514, no. 7520 (October 2014): 54–58.

138 *the presence of certain "helper" clones may also support the cancer cell colony* Marusyk et al., "Non-Cell-Autonomous Driving of Tumour Growth Supports Sub-Clonal Heterogeneity."

138 *The volatility of the environment—in terms of foraging success, the possibility of illness and injury, the chance of natural disasters, and extreme weather—make it hard to survive as an individual human* Cronk et al., "Managing Risk through Cooperation."

138 *social insects like honeybees and ants have evolved to live in large colonies that help buffer them from the harsh environments they live in, in addition to enabling division of labor on a massive scale* Hölldobler and Wilson, *The Superorganism*.

138 *Fitness interdependence in humans often happens in situations where the survival or reproductive success of individuals is yoked together (such as mating relationships in which the survival and success of mutual offspring is at play), in periods of war (when soldiers are dependent on one another for survival), and in harsh and unpredictable environments (where it may be impossible to survive as a loner)* A. Aktipis et al., "Understanding Cooperation through Fitness Interdependence," *Nature Human Behavior* 2 (2018): 429–431.

139 *Some cooperation theorists categorically assert that nothing can evolve "for the good of the group," including cancer cells* Archetti and Pienta, "Cooperation among Cancer Cells: Applying Game Theory to Cancer," *Nature Reviews Cancer* 19, no. 2 (February 2019): 110–17.

140 *neither of them matches current data about what metastasis actually looks like*   S. Turajlic and C. Swanton, "Metastasis as an Evolutionary Process," *Science* 352 (2016): 169–175.

140 *cancer proceeds in stages as each mutation builds on the last one in a step-by-step fashion*   Nowell, "The Clonal Evolution of Tumor Cell Populations."

141 *There are elements of both models in the data—even within one sample of one cancer*   Turajlic and Swanton, "Metastasis as an Evolutionary Process."

142 *Another phenomenon that does not fit with either model is tumor reseeding, in which cells from metastases appear back in the primary tumor*   Turajlic and Swanton, "Metastasis as an Evolutionary Process."

142 *This concept is similar to the haystack model, a classic model of the evolution of cooperation via multilevel selection*   John Maynard Smith, "Group Selection and Kin Selection," *Nature* 201 (March 1964): 1145.

142 *cancer cells are capable of conditional movement, making it even more likely that selection will favor cooperation within these cellular groups*   C. A. Aktipis, "Is Cooperation Viable in Mobile Organisms? Simple Walk Away Rule Favors the Evolution of Cooperation in Groups," *Evolution and Human Behavior* 32, no. 4 (2011): 263–76; Joshua D. Schiffman, Richard M. White, Trevor A. Graham, Qihong Huang, and Athena Aktipis, "The Darwinian Dynamics of Motility and Metastasis," in *Frontiers in Cancer Research* (New York: Springer, 2016), 135–76.

143 *Do cancer cell colonies vary? Indeed they do. During metastasis, colonies of cancer cells vary genetically.*   Marco Gerlinger et al., "Genomic Architecture and Evolution of Clear Cell Renal Cell Carcinomas Defined by Multiregion Sequencing," *Nature Genetics* 46, no. 3 (March 2014): 225–33; M. Gerlinger et al., "Intratumor Heterogeneity and Branched Evolution Revealed by Multiregion Sequencing," *New England Journal of Medicine* 366, no. 10 (2012): 883–92.

143 *the cancer cell colonies have different rates of survival and creation of new colonies*   Samra Turajlic and Charles Swanton, "Metastasis as an Evolutionary Process," *Science* 352, no. 6282 (April 2016): 169–75.

143 *some cell colonies give rise to many new colonies, whereas other colonies do not appear to spawn new ones*   Turajlic and Swanton, "Metastasis as an Evolutionary Process."

143 *tiny metastases will grow rapidly because the primary tumor is no longer monopolizing nutrients and producing inhibitory factors*   Paula Chiarella et al., "Concomitant Tumor Resistance," *Cancer Letters* 324, no. 2 (November 2012): 133–41.

143 *This phenomenon of primary tumors suppressing metastases, known as "concomitant tumor resistance," has been widely observed in both animal experiments and in human patients*   Chiarella et al., "Concomitant Tumor Resistance."

144 *clusters of cancer cells often colonize together—and with greater success than individual cells*   Aceto et al., "Circulating Tumor Cell Clusters Are Oligoclonal Precursors of Breast Cancer Metastasis."

144   *clusters of cancer cells evolve to have some reproductive division of labor, with some cells proliferating while other cells support the proliferating cells, almost like a protomulticellular organism*   Sprouffske et al., "An Evolutionary Explanation for the Presence of Cancer Nonstem Cells in Neoplasms."

144   *life cycles with distinct phases of growth and reproduction, as well as life history strategies*   Joshua D. Schiffman et al., "The Darwinian Dynamics of Motility and Metastasis," in *Frontiers in Cancer Research* (New York: Springer, 2016), 135–76.

145   *She rightly points out that most incipient metastases fail before they even have a chance to take hold, getting destroyed in the circulatory system or simply failing to colonize*   Lean and Plutynski, "The Evolution of Failure."

145   *multilevel selection is not relevant to late-stage cancer because metastases are unlikely to "reproduce" in a way that reliably leads to heritability of variation among them*   Pierre-Luc Germain and Lucie Laplane, "Metastasis as Supra-Cellular Selection? A Reply to Lean and Plutynski," *Biology and Philosophy* 32, no. 2 (March 2017): 281–87.

145   *there is evidence for heritability of traits at the colony level*   Alison A. Bockoven, Shawn M. Wilder, and Micky D. Eubanks, "Intraspecific Variation among Social Insect Colonies: Persistent Regional and Colony-Level Differences in Fire Ant Foraging Behavior," *PloS One* 10, no. 7 (2015): e0133868; Justin T. Walsh, Simon Garnier, and Timothy A. Linksvayer, "Ant Collective Behavior Is Heritable and Shaped by Selection," *bioRxiv* (March 2019): 567503.

146   *a process called early dissemination*   Hedayatollah Hosseini et al., "Early Dissemination Seeds Metastasis in Breast Cancer," *Nature* 540, no. 7634 (December 2016): 552–58.

146   *We also know that large tumors can effectively suppress the growth of smaller tumors*   Chiarella et al., "Concomitant Tumor Resistance."

146   *during metastasis, a metapopulation structure emerges that could select for cell colonies that are good at metastasizing*   Schiffman et al., "The Darwinian Dynamics of Motility and Metastasis," 135–76.

147   *if selection among metastatic colonies is the mechanism that pushes forward cancer, then removing the primary tumor would not stop a metastatic cascade (and, in fact, there is evidence that removing a primary tumor can sometimes harm the patient)*   Chiarella et al., "Concomitant Tumor Resistance."

148   *if two populations of cancer cells with different mutations (say, one that produces a growth factor and another that produces a factor that allows for invasion) happen to be near each other, they may end up providing benefits for one other*   Axelrod, Axelrod, and Pienta, "Evolution of Cooperation among Tumor Cells."

148   *In harsh environments where it is difficult to survive without producing these kinds of factors, by-product mutualism becomes even more likely because "cheating" is not really a viable option*   Dugatkin, "Animal Cooperation among Unrelated Individuals."

148   *These kinds of by-product benefits are likely a part of the explanation for why we see cancer cells cooperating inside tumors*   Axelrod, Axelrod, and Pienta, "Evolution of Cooperation among Tumor Cells."

148 *the cells that provide these benefits for one another end up interacting preferentially as a result of spatial proximity or other factors that promote positive assortment* Fletcher and Doebeli, "A Simple and General Explanation for the Evolution of Altruism."

150 *cancer cells can use cooperation and coordination to invade tissues; for example, they can use electrical and chemical signaling to move together as a small group or even as a long conga line of cells through tissues and membranes* Tai, Tai, and Zhao, "Electrically Stimulated Cell Migration and Its Contribution to Wound Healing"; Haeger et al., "Collective Cell Migration."

150 *when clusters of cancer cells colonize in groups, they are more successful than when they are alone* Aceto et al., "Circulating Tumor Cell Clusters Are Oligoclonal Precursors of Breast Cancer Metastasis."

150 *We can think of these microbes as cooperating with us, their multicellular hosts, in ways that are beneficial for both us and them* H. Wasielewski, J. Alcock, and A. Aktipis, "Resource Conflict and Cooperation between Human Host and Gut Microbiota: Implications for Nutrition and Health," *Annals of the New York Academy of Sciences* 1372, no. 1 (2016): 20–28.

151 *They thrive by exploiting us—using our resources for their own survival and proliferation* Wasielewski, Alcock, and Aktipis, "Resource Conflict and Cooperation between Human Host and Gut Microbiota."

151 *About 10 to 20 percent of human cancers are associated with specific microbial species* Catherine de Martel et al., "Global Burden of Cancers Attributable to Infections in 2008: A Review and Synthetic Analysis," *Lancet Oncology* 13, no. 6 (June 2012): 607–15.

151 *and many other microbes (as well as multicellular parasites) are suspected to play a role in cancer risk, even if it is indirect* Paul W. Ewald, "An Evolutionary Perspective on Parasitism as a Cause of Cancer," in *Advances in Parasitology*, vol. 68 (Cambridge, MA: Academic Press, 2009), 21–43.

151 *many cancers in wildlife are associated with microbial infections* Patricia A. Pesavento et al., "Cancer in Wildlife: Patterns of Emergence," *Nature Reviews Cancer* 18, no. 10 (October 2018): 646–61.

151 *Cancer cells and microbes can cooperate with one another, teaming up in order to better exploit the multicellular body* C. Whisner and A. Aktipis, "The Role of the Microbiome in Cancer Initiation and Progression: How Microbes and Cancer Cells Utilize Excess Energy and Promote One Another's Growth," *Current Nutrition Reports* 8, no. 1 (March 2019): 42–51.

151 *This multispecies cooperation can evolve simply because of assortment—the preferential interaction of cooperators with one another* Fletcher and Doebeli, "A Simple and General Explanation for the Evolution of Altruism."

151 *Some microbes, like the virus HPV, are quite direct about it: they get into the nucleus of the cell and increase cell proliferation, which increases the risk of cancer, in part by interfering with the p53 protein* Nubia Muñoz et al., "Chapter 1: HPV in the Etiology of Human Cancer," *Vaccine* 24, suppl. 3 (August 2006): S3/1–10.

151 *There are more subtle ways that microbes can increase cancer risk, including producing genotoxins that damage DNA* Jean-Philippe Nougayrède et al.,

"Escherichia Coli Induces DNA Double-Strand Breaks in Eukaryotic Cells," *Science* 313, no. 5788 (August 2006): 848–51; Aadra P. Bhatt, Matthew R. Redinbo, and Scott J. Bultman, "The Role of the Microbiome in Cancer Development and Therapy," *CA: A Cancer Journal for Clinicians* 67, no. 4 (July 2017): 326–44; Andrew C. Goodwin et al., "Polyamine Catabolism Contributes to Enterotoxigenic Bacteroides Fragilis-Induced Colon Tumorigenesis," *Proceedings of the National Academy of Sciences of the United States of America* 108, no. 37 (September 2011): 15354–59.

151   *and producing virulence factors that increase cell proliferation*   Antony Cougnoux et al., "Bacterial Genotoxin Colibactin Promotes Colon Tumour Growth by Inducing a Senescence-Associated Secretory Phenotype," *Gut* 63, no. 12 (December 2014): 1932–42; Guillaume Dalmasso et al., "The Bacterial Genotoxin Colibactin Promotes Colon Tumor Growth by Modifying the Tumor Microenvironment," *Gut Microbes* 5, no. 5 (2014): 675–80.

151   *Microbes and cancer cells can also produce growth factors for one another*   Cougnoux et al., "Bacterial Genotoxin Colibactin Promotes Colon Tumour Growth by Inducing a Senescence-Associated Secretory Phenotype"; Dalmasso et al., "The Bacterial Genotoxin Colibactin Promotes Colon Tumor Growth by Modifying the Tumor Microenvironment."

151   *and have the ability to protect one another from the immune system*   T. Hussell et al., "Helicobacter Pylori-Specific Tumour-Infiltrating T Cells Provide Contact Dependent Help for the Growth of Malignant B Cells in Low-Grade Gastric Lymphoma of Mucosa-Associated Lymphoid Tissue," *Journal of Pathology* 178, no. 2 (February 1996): 122–27; Marc Lecuit et al., "Immunoproliferative Small Intestinal Disease Associated with Campylobacter Jejuni," *New England Journal of Medicine* 350, no. 3 (January 2004): 239–48; Andrés J. M. Ferreri et al., "Chlamydophila Psittaci Eradication with Doxycycline as First-Line Targeted Therapy for Ocular Adnexae Lymphoma: Final Results of an International Phase II Trial," *Journal of Clinical Oncology* 30, no. 24 (August 2012): 2988–94; Brian Goodman and Humphrey Gardner, "The Microbiome and Cancer," *Journal of Pathology* 244, no. 5 (April 2018): 667–76; Shaoguang Wu et al., "A Human Colonic Commensal Promotes Colon Tumorigenesis via Activation of T Helper Type 17 T Cell Responses," *Nature Medicine* 15, no. 9 (September 2009): 1016–22.

152   *microbes can also help cancer cells to invade and metastasize by producing toxins that transform cancer cells from more sedentary to more motile*   Sara Gaines et al., "How the Microbiome Is Shaping Our Understanding of Cancer Biology and Its Treatment," *Seminars in Colon and Rectal Surgery* 29, no. 1 (March 2018): 12–16.

152   *and producing quorum sensing molecules that contribute to metastasis*   Evelien Wynendaele et al., "Crosstalk between the Microbiome and Cancer Cells by Quorum Sensing Peptides," *Peptides* 64 (February 2015): 40–48.

152   *Doctors use the bacterium* Mycobacterium bovis *BCG in the treatment of bladder cancer*   A. M. Chakrabarty, "Microorganisms and Cancer: Quest for a Therapy," *Journal of Bacteriology* 185, no. 9 (May 2003): 2683–86.

152 *There are many different ways that microbes and their products can help treat cancer, including activating the immune system, inducing cell death, and inhibiting new blood vessel growth* Chakrabarty, "Microorganisms and Cancer."

152 *experiments have found that mice with their commensal microbiomes intact have better responses to therapy than mice that have been given antibiotics* Noriho Iida et al., "Commensal Bacteria Control Cancer Response to Therapy by Modulating the Tumor Microenvironment," *Science* 342, no. 6161 (November 2013): 967–70.

152 *Microbes can enhance intestinal barrier function, improve immune function, inhibit cell proliferation, and help regulate metabolism* Alasdair J. Scott et al., "Pre-, Pro- and Synbiotics in Cancer Prevention and Treatment—A Review of Basic and Clinical Research," *ecancermedicalscience* 12 (September 2018): 869.

152 *a meta-analysis found that consumption of a lot of fiber (which is a prebiotic, since it feeds beneficial microbes) was associated with lower colon cancer risk* Qiwen Ben et al., "Dietary Fiber Intake Reduces Risk for Colorectal Adenoma: A Meta-Analysis," *Gastroenterology* 146, no. 3 (March 2014): 689–99.

152 *research in this area is new and not all studies find a protective effect, but the finding is intriguing* Scott et al., "Pre-, Pro- and Synbiotics in Cancer Prevention and Treatment."

152 *they can turn on virulence genes in response to low nutrients* Wasielewski, Alcock, and Aktipis, "Resource Conflict and Cooperation between Human Host and Gut Microbiota."

152 *in some cases cooperation occurs between cancer cells and microbes during progression* Whisner and Aktipis, "The Role of the Microbiome in Cancer Initiation and Progression."

154 *mutations in the gene coding for the NOTCH1 receptor (an intercellular signaling protein involved in many aspects of cell function) occur in about 10 percent of esophageal cancers, and so researchers had assumed that these mutations probably contributed to the cancer* Inigo Martincorena et al., "Somatic Mutant Clones Colonize the Human Esophagus with Age," *Science* 362, no. 6417 (November 2018): 911–17.

154 *they discovered that mutations in NOTCH1 were far more common—present in 30 to 80 percent of the normal esophageal tissue compared to the 10 percent in cancerous esophageal tissues found in previous studies* Martincorena et al., "Somatic Mutant Clones Colonize the Human Esophagus with Age."

154 *This finding, that NOTCH1 mutations are more strongly associated with normal esophageal tissue than with esophageal cancer, has been replicated* Akira Yokoyama et al., "Age-Related Remodelling of Oesophageal Epithelia by Mutated Cancer Drivers," *Nature* 565, no. 7739 (January 2019): 312–17.

155 *this might be part of the "program" that multicellular bodies evolved to minimize cancer risk* Kelly C. Higa and James DeGregori, "Decoy Fitness Peaks, Tumor Suppression, and Aging," *Aging Cell* 18, no. 3 (June 2019): e12938.

155   *One mechanism could be mutation "hotspots"*   Igor B. Rogozin and Youri I. Pavlov, "Theoretical Analysis of Mutation Hotspots and Their DNA Sequence Context Specificity," *Mutation Research* 544, no. 1 (September 2003): 65–85.

155   *Mutations can be induced by cell-level stresses like DNA damage*   Raul Correa et al., "Oxygen and RNA in Stress-Induced Mutation," *Current Genetics* 64, no. 4 (August 2018): 769–76; S. M. Rosenberg, "Evolving Responsively: Adaptive Mutation," *Nature Reviews Genetics* 2, no. 7 (July 2001): 504–15.

155   *some microbes can be beneficial to humans simply because they can occupy the ecological space in and on our bodies so that disease-causing microbes can't*   Wasielewski, Alcock, and Aktipis, "Resource Conflict and Cooperation between Human Host and Gut Microbiota."

156   *genes inside the genome can sometimes work at cross-purposes to one another, promoting their own replication at the expense of the cell or changing the expression state of the cell in ways that improve the gene's fitness*   J. Featherston and P. M. Durand, "Cooperation and Conflict in Cancer: An Evolutionary Perspective," *South African Journal of Science* 108, no. 9–10 (January 2012).

156   *This was one of the critical transitions during the evolution of life—it allowed for cooperation and coordination of genes within a genome, allowing cells to develop complicated behaviors that free-floating pieces of DNA never could*   Buss, *The Evolution of Individuality*; John Maynard Smith and Eörs Szathmáry, *The Major Transitions in Evolution* (Oxford: Oxford University Press, 1995).

157   *mobile genetic elements "are functionally analogous to the presumed ancient replicators that cooperated to form primitive protein coding genomes"*   Featherston and Durand, "Cooperation and Conflict in Cancer."

157   *disruptions in the epigenetics of cancer cell genomes cause disruptions in the normal control of these mobile genetic elements, which may cause further genomic alterations as they copy themselves around the cancer genomes*   Kathleen H. Burns, "Transposable Elements in Cancer," *Nature Reviews Cancer* 17, no. 7 (July 2017): 415–24.

157   *mobile elements can cause genome damage and "dysregulation of genome replication or cell cycling [and] disruption of cooperative cellular behaviour," and that disruption of normal expression in areas of the genome that harbor mobile genetic elements is prevalent across many cancers*   Featherston and Durand, "Cooperation and Conflict in Cancer."

158   *If the DNA is outside the chromosomes, its extrachromosomal existence implies that it has already escaped from genome-level controls of DNA replication and may be "free to proliferate or indulge in selfish behaviour," according to Featherston and Durand*   Featherston and Durand, "Cooperation and Conflict in Cancer."

158   *And this extrachromosomal DNA contained extra copies of driver oncogenes (genes associated with cancer), suggesting that they may be playing a causal role in cancer rather than just being associated with the disease*   Kristen M. Turner et al., "Extrachromosomal Oncogene Amplification Drives Tumour Evolution and Genetic Heterogeneity," *Nature* 543 (February 2017): 122.

Chapter 7. How to Control Cancer

160   *In 1972, just one year after signing the National Cancer Act, Richard Nixon signed an act creating a new national policy around an agricultural approach called "integrated pest management"*   T. D. Landis and R. K. Dumroese, "Integrated Pest Management—An Overview and Update," *Forest Nursery Notes* (2014), https://www.researchgate.net/profile/R_Kasten_Dumroese /publication/272682105_Integrated_pest_management-an_overview_and _update/links/54ebbce10cf2082851be7e2b.pdf.

161   *The next strategy is to reduce the numbers of those pests, applying treatment to bring them back below the threshold where they are not doing too much damage*   D. G. Alston, *The Integrated Pest Management (IPM) Concept* (Logan: Utah State University Extension and Utah Plant Pest Diagnostic Laboratory, 2011).

162   *The evolution of resistance to chemotherapy has been a problem for every kind of drug that has ever been tried, including targeted therapies like EGFR blockades and HER2-targeted therapies*   Luis A. Diaz Jr. et al., "The Molecular Evolution of Acquired Resistance to Targeted EGFR Blockade in Colorectal Cancers," *Nature* 486, no. 7404 (June 2012): 537–40; Rita Nahta et al., "Mechanisms of Disease: Understanding Resistance to HER2-Targeted Therapy in Human Breast Cancer," *Nature Clinical Practice Oncology* 3, no. 5 (May 2006): 269–80; Robert A. Gatenby et al., "Adaptive Therapy," *Cancer Research* 69, no. 11 (June 2009): 4894–903.

164   *adaptive therapy kept the mouse tumors under control*   Gatenby et al., "Adaptive Therapy."

164   *the adaptive therapy approach allowed these mice to "survive indefinitely with a small, reasonably stable tumor burden"*   Gatenby et al., "Adaptive Therapy."

164   *tumors could be controlled with a smaller and smaller dose as time went on*   Pedro M. Enriquez-Navas et al., "Exploiting Evolutionary Principles to Prolong Tumor Control in Preclinical Models of Breast Cancer," *Science Translational Medicine* 8, no. 327 (February 2016): 327ra24.

164   *more stable environments can select for cells that have slower life history strategies*   Aktipis et al., "Life History Trade-Offs in Cancer Evolution."

165   *As of October 2017, when Zhang and Gatenby's pilot study was accepted for publication, only one of the eleven patients' cancers had progressed*   Jingsong Zhang et al., "Integrating Evolutionary Dynamics into Treatment of Metastatic Castrate-Resistant Prostate Cancer," *Nature Communications* 8, no. 1 (November 2017): 1816.

166   *Some five-year survival rates are extremely high, near 100 percent for early-stage thyroid cancer and between 60 and 85 percent for childhood leukemias, depending on the type, according to the American Cancer Society.*   R. L. Siegel, K. D. Miller, and A. Jemal, "Cancer Statistics, 2018," *CA: A Cancer Journal for Clinicians* 68, no. 1 (2018): 7–30.

167   *some studies suggest that patients can do just as well with palliative care (which is focused on improving patient quality of life and reducing pain) as with expensive*

and painful treatment that is aimed at curing the cancer   Jennifer S. Temel
et al., "Early Palliative Care for Patients with Metastatic Non-Small-Cell
Lung Cancer," *New England Journal of Medicine* 363, no. 8 (August 2010):
733–42; Stephen R. Connor et al., "Comparing Hospice and Nonhospice
Patient Survival among Patients Who Die within a Three-Year Window,"
Journal of Pain and Symptom Management 33, no. 3 (March 2007): 238–46.

167   *"Drug use causes drug resistance, a firestorm of drugs removes the competitors
of the very things we fear: the cells and bugs we can't kill"*   Andrew F. Read,
"The Selfish Germ," *PLoS Biology* 15, no. 7 (July 2017): e2003250.

167   *We established a set of principles for how to measure cancer's evolvability,
called the Evo-Index and Eco-Index*   Carlo C. Maley et al., "Classifying the
Evolutionary and Ecological Features of Neoplasms," *Nature Reviews Cancer*
17, no. 10 (October 2017): 605–19.

172   *it often takes decades from the first mutations to finding cancer*   S. Jones et al.,
"Comparative Lesion Sequencing Provides Insights into Tumor Evolution,"
Proceedings of the National Academy of Sciences of the United States of America
105, no. 11 (2008): 4283–88.

172   *one baby aspirin per day helped reduce the mutation rate by an order of
magnitude*   Rumen L. Kostadinov et al., "NSAIDs Modulate Clonal Evolution
in Barrett's Esophagus," *PLoS Genetics* 9, no. 6 (June 2013): e1003553.

172   *multiple studies have shown that NSAIDs slow progression to esophageal cancer
as well as many other cancers*   Jack Cuzick et al., "Aspirin and Non-Steroidal
Anti-Inflammatory Drugs for Cancer Prevention: An International Consensus
Statement," *Lancet Oncology* 10, no. 5 (May 2009): 501–7; Enrico Flossmann,
Peter M. Rothwell, and British Doctors Aspirin Trial and the U.K.-TIA
Aspirin Trial, "Effect of Aspirin on Long-Term Risk of Colorectal Cancer:
Consistent Evidence from Randomised and Observational Studies," *Lancet*
369, no. 9573 (May 2007): 1603–13; Peter M. Rothwell et al., "Effect of Daily
Aspirin on Long-Term Risk of Death due to Cancer: Analysis of Individual
Patient Data from Randomised Trials," *Lancet* 377, no. 9759 (January 2011):
31–41; Thomas L. Vaughan et al., "Non-Steroidal Anti-Inflammatory Drugs
and Risk of Neoplastic Progression in Barrett's Oesophagus: A Prospective
Study," *Lancet Oncology* 6, no. 12 (December 2005): 945–52, https://doi.org
/10.1016/S1470-2045(05)70431-9.

172   *This may be because NSAIDs reduce the mutation rate directly*   Kostadinov
et al., "NSAIDs Modulate Clonal Evolution in Barrett's Esophagus."

173   *Gatenby and his colleagues found that they could decrease cell proliferation
of resistant cells in Petri dishes and that the growth rate of resistant cell lines
(compared to similar nonresistant cell lines) was lower in a mouse model with
the administration of the ersatzdroges*   Kam et al., "Sweat but No Gain,"
International Journal of Cancer 136, no. 4 (2015): E188–96.

174   *The size of the primary tumor was not affected, but by returning the tumor
environment to a more pH neutral state, the metastases were significantly
decreased—and this led to an improvement in survival for the mice receiving
what Gatenby's team called "bicarbonate therapy"*   Ian F. Robey et al.,

"Bicarbonate Increases Tumor pH and Inhibits Spontaneous Metastases," *Cancer Research* 69, no. 6 (March 2009): 2260–68.

174 *When oxygen levels are low, cancer cells are more likely to invade and metastasize* Erinn B. Rankin and Amato J. Giaccia, "Hypoxic Control of Metastasis," *Science* 352, no. 6282 (April 2016): 175–80.

174 *Studies suggest that normalizing the resource delivery to the tumor can actually reduce metastasis* M. Mazzone et al., "Heterozygous Deficiency of PHD2 Restores Tumor Oxygenation and Inhibits Metastasis via Endothelial Normalization," *Cell* 136, no. 5 (2009): 839–51.

174 *using low levels of antiangiogenic drugs (which help to regulate blood flow to the tumor) can improve response to treatments* Yuhui Huang et al., "Vascular Normalization as an Emerging Strategy to Enhance Cancer Immunotherapy," *Cancer Research* 73, no. 10 (May 2013): 2943–48.

175 *It's similar to what Winston Churchill said about democracy: that it is the worst form of government except for all those other forms that have been tried.* Winston S. Churchill, November 11, 1947, The International Churchill Society, https://winstonchurchill.org/resources/quotes/the-worst-form-of-government/.

175 *Many cancers have cells with mutations in* TP53 Karen H. Vousden and Xin Lu, "Live or Let Die: The Cell's Response to p53," *Nature Reviews Cancer* 2, no. 8 (August 2002): 594–604.

176 *Some potential strategies for rebooting cellular self-control are restoring* TP53 *function when it is lost* A. N. Bullock and A. R. Fersht, "Rescuing the Function of Mutant p53," *Nature Reviews Cancer* 1, no. 1 (October 2001): 68–76.

176 *elephant* TP53 *can restore normal p53 function and apoptosis in human osteoscarcoma cells* Lisa M. Abegglen et al., "Abstract A25: Elephant p53 (EP53) Enhances and Restores p53-Mediated Apoptosis in Human and Canine Osteosarcoma," *Clinical Cancer Research* 24, no. 2 suppl. (January 2018): 48–49.

177 *reducing inflammation helps reduce the risk of cancer* Kostadinov et al., "NSAIDs Modulate Clonal Evolution in Barrett's Esophagus."

179 *By restoring the immune system's ability to detect cellular cheaters, immune checkpoint blockade therapies have been successful at treating previously intractable cancers, including melanomas and lung cancers, in some patients* Drew M. Pardoll, "The Blockade of Immune Checkpoints in Cancer Immunotherapy," *Nature Reviews Cancer* 12, no. 4 (March 2012): 252–64; Suzanne L. Topalian et al., "Safety, Activity, and Immune Correlates of Anti-PD-1 Antibody in Cancer," *New England Journal of Medicine* 366, no. 26 (June 2012): 2443–54; F. Stephen Hodi et al., "Improved Survival with Ipilimumab in Patients with Metastatic Melanoma," *New England Journal of Medicine* 363, no. 8 (August 2010): 711–23.

179 *cancer cells do still evolve resistance to immunotherapies* Russell W. Jenkins et al., "Mechanisms of Resistance to Immune Checkpoint Inhibitors," *British Journal of Cancer* 118, no. 1 (January 2018): 9–16.

179 *interfering with adhesion in circulating cell clusters to hopefully reduce the likelihood of metastasis (since cell clusters have been found to be more likely to*

metastasize than single cells)   Aceto et al., "Circulating Tumor Cell Clusters Are Oligoclonal Precursors of Breast Cancer Metastasis."

179   *Disruption to cooperative cellular signaling among cancer cells is another potential strategy for cancer control*   G. Jansen, R. Gatenby, and C. A. Aktipis, "Opinion: Control vs. Eradication; Applying Infectious Disease Treatment Strategies to Cancer," *Proceedings of the National Academy of Sciences of the United States of America* 112, no. 4 (2015): 937–38.

180   *Perhaps we should be searching for and using drugs that disrupt cell cooperation by interfering with cancer cell communication*   Jansen, Gatenby, and Aktipis, "Opinion: Control vs. Eradication."

180   *Higher levels of plakoglobins are associated with worse patient outcomes*   Aceto et al., "Circulating Tumor Cell Clusters Are Oligoclonal Precursors of Breast Cancer Metastasis."

181   *Interfering with public good production is one example of an intervention targeted at disrupting cancer cell cooperation*   John W. Pepper, "Drugs That Target Pathogen Public Goods Are Robust against Evolved Drug Resistance," *Evolutionary Applications* 5, no. 7 (November 2012): 757–61.

182   *the late evolutionary biologist Stephen Jay Gould said, "I prefer the more martial view that death is the ultimate enemy—and I find nothing reproachable in those who rage mightily against the dying of the light."*   Stephen J. Gould, "The Median Isn't the Message," *Discover* 6, no. 6 (1985): 40–42.

183   *Many researchers in the cancer community are working to identify the key parameters that should be used when we're making strategic decisions about how to treat cancer*   Gatenby et al., "Adaptive Therapy"; Maley et al., "Classifying the Evolutionary and Ecological Features of Neoplasms."; Elsa Hansen, Robert J. Woods, and Andrew F. Read, "How to Use a Chemotherapeutic Agent When Resistance to It Threatens the Patient," *PLoS Biology* 15, no. 2 (2017): e2001110.

183   *This shift in our thinking allows us all to consider cancer as a chronic and manageable disease, which opens up new pathways for treating and preventing cancer*   Robert A. Gatenby, "A Change of Strategy in the War on Cancer," *Nature* 459, no. 7246 (2009): 508–9; Sui Huang, "The War on Cancer: Lessons from the War on Terror," *Frontiers in Oncology* 4 (October 2014): 293; Bryan Oronsky et al., "The War on Cancer: A Military Perspective," *Frontiers in Oncology* 4 (2014): 387.

BIBLIOGRAPHY

Abegglen, Lisa M., Aleah F. Caulin, Ashley Chan, Kristy Lee, Rosann Robinson, Michael S. Campbell, Wendy K. Kiso, et al. "Potential Mechanisms for Cancer Resistance in Elephants and Comparative Cellular Response to DNA Damage in Humans." *Journal of the American Medical Association* 314, no. 17 (November 2015): 1850–60.

Abegglen, Lisa M., Cristhian Toruno, Lauren N. Donovan, Rosann Robinson, Mor Goldfeder, Genevieve Couldwell, Wendy K. Kiso, et al. "Abstract A25: Elephant p53 (EP53) Enhances and Restores p53-Mediated Apoptosis in Human and Canine Osteosarcoma." *Clinical Cancer Research* 24, no. 2 suppl. (January 2018): 48–49.

Aceto, Nicola, Aditya Bardia, David T. Miyamoto, Maria C. Donaldson, Ben S. Wittner, Joel A. Spencer, Min Yu, Adam Pely, Amanda Engstrom, and Huili Zhu. "Circulating Tumor Cell Clusters Are Oligoclonal Precursors of Breast Cancer Metastasis." *Cell* 158, no. 5 (2014): 1110–22.

Agrios, G. N. *Plant Pathology.* Boston: Elsevier Academic Press, 2005.

Aktipis, A. "Principles of Cooperation across Systems: From Human Sharing to Multicellularity and Cancer." *Evolutionary Applications* 9, no. 1 (2015): 17–36.

Aktipis, A., L. Cronk, D. Sznycer, J. Alcock, J. Ayers, C. Baciu, D. Balliet, et al. "Understanding Cooperation through Fitness Interdependence." *Nature Human Behavior* 2 (2018): 429–431.

Aktipis, C. A. "Is Cooperation Viable in Mobile Organisms? Simple Walk Away Rule Favors the Evolution of Cooperation in Groups." *Evolution and Human Behavior* 32, no. 4 (2011): 263–76.

Aktipis, C. A. "Know When to Walk Away: Contingent Movement and the Evolution of Cooperation." *Journal of Theoretical Biology* 231, no. 2 (2004): 249–60.

Aktipis, C. A., A. M. Boddy, R. A. Gatenby, J. S. Brown, and C. C. Maley. "Life History Trade-Offs in Cancer Evolution." *Nature Reviews Cancer* 13, no. 12 (2013): 883–92.

Aktipis, C. A., Amy M. Boddy, G. Jansen, U. Hibner, M. E. Hochberg, C. C. Maley, and G. S. Wilkinson. "Cancer across the Tree of Life: Cooperation and Cheating in Multicellularity." *Philosophical Transactions of the Royal Society of London, Series B: Biological Sciences* 370, no. 1673 (2015). https://doi.org/10.1098/rstb.2014.0219.

Aktipis, C. A., and E. Fernandez-Duque. "Parental Investment without Kin Recognition: Simple Conditional Rules for Parent–Offspring Behavior." *Behavioral Ecology and Sociobiology* 65, no. 5 (May 2011): 1079–91.

Aktipis, C. A., C. C. Maley, and J. W. Pepper. "Dispersal Evolution in Neoplasms: The Role of Disregulated Metabolism in the Evolution of Cell Motility." *Cancer Prevention Research* 5, no. 2 (2012): 266–75.

Aktipis, C. A., and R. M. Nesse. "Evolutionary Foundations for Cancer Biology." *Evolutionary Applications* 6, no. 1 (2013): 144–59.

Aktipis, C. Athena, Bruce J. Ellis, Katherine K. Nishimura, and Robert A. Hiatt. "Modern Reproductive Patterns Associated with Estrogen Receptor Positive but Not Negative Breast Cancer Susceptibility." *Evolution, Medicine, and Public Health* 2015, no. 1 (2015): 52–74. https://dx.doi.org/10.1093/emph/eou028.

Allam, Bassem, and David Raftos. "Immune Responses to Infectious Diseases in Bivalves." *Journal of Invertebrate Pathology* 131 (October 2015): 121–36.

Alston, D. G. *The Integrated Pest Management (IPM) Concept*. Logan: Utah State University Extension and Utah Plant Pest Diagnostic Laboratory, 2011.

Alvarado, L. C. "Do Evolutionary Life-History Trade-Offs Influence Prostate Cancer Risk? A Review of Population Variation in Testosterone Levels and Prostate Cancer Disparities." *Evolutionary Applications* 6, no. 1 (2013): 117–33.

Archetti, Marco. "Cooperation between Cancer Cells." *Evolution, Medicine, and Public Health* 2018, no. 1 (January 2018): 1.

Archetti, Marco, and Kenneth J. Pienta. "Cooperation among Cancer Cells: Applying Game Theory to Cancer." *Nature Reviews Cancer* 19, no. 2 (February 2019): 110–17.

Axelrod, R., D. E. Axelrod, and K. J. Pienta. "Evolution of Cooperation among Tumor Cells." *Proceedings of the National Academy of Sciences of the United States of America* 103, no. 36 (2006): 13474–79.

Axelrod, Robert, and W. D. Hamilton. "The Evolution of Cooperation." *Science* 211, no. 4489 (1981): 1390–96.

Bagnardi, V., M. Rota, E. Botteri, I. Tramacere, F. Islami, V. Fedirko, L. Scotti, et al. "Alcohol Consumption and Site-Specific Cancer Risk: A Comprehensive Dose-Response Meta-Analysis." *British Journal of Cancer* 112, no. 3 (February 2015): 580–93.

Belov, Katherine. "The Role of the Major Histocompatibility Complex in the Spread of Contagious Cancers." *Mammalian Genome* 22, no. 1–2 (February 2011): 83–90.

Ben, Qiwen, Yunwei Sun, Rui Chai, Aihua Qian, Bin Xu, and Yaozong Yuan. "Dietary Fiber Intake Reduces Risk for Colorectal Adenoma: A Meta-Analysis." *Gastroenterology* 146, no. 3 (March 2014): 689–99.

Bergmüller, Ralph, Rufus A. Johnstone, Andrew F. Russell, and Redouan Bshary. "Integrating Cooperative Breeding into Theoretical Concepts of Cooperation." *Behavioural Processes* 76, no. 2 (2007): 61–72.

Bhatt, Aadra P., Matthew R. Redinbo, and Scott J. Bultman. "The Role of the Microbiome in Cancer Development and Therapy." *CA: A Cancer Journal for Clinicians* 67, no. 4 (July 2017): 326–44.

Bissell, M. J., and W. C. Hines. "Why Don't We Get More Cancer? A Proposed Role of the Microenvironment in Restraining Cancer Progression." *Nature Medicine* 17, no. 3 (2011): 320–29.

Bissell, M. J., and D. Radisky. "Putting Tumours in Context." *Nature Reviews Cancer* 1, no. 1 (2001): 46–54.

Bockoven, Alison A., Shawn M. Wilder, and Micky D. Eubanks. "Intraspecific Variation among Social Insect Colonies: Persistent Regional and Colony-Level Differences in Fire Ant Foraging Behavior." *PloS One* 10, no. 7 (2015): e0133868.

Boddy, A. M., H. Kokko, F. Breden, G. S. Wilkinson, and C. A. Aktipis. "Cancer Susceptibility and Reproductive Trade-Offs: A Model of the Evolution of Cancer Defences." *Philosophical Transactions of the Royal Society of London, Series B: Biological Sciences* 370, no. 1673 (2015). https://doi.org/10.1098/rstb.2014.0220.

Bonner, John Tyler. "The Origins of Multicellularity." *Integrative Biology: Issues, News, and Reviews* 1, no. 1 (1998): 27–36.

Bouvard, Véronique, Dana Loomis, Kathryn Z. Guyton, Yann Grosse, Fatiha El Ghissassi, Lamia Benbrahim-Tallaa, Neela Guha, Heidi Mattock, Kurt Straif, and International Agency for Research on Cancer Monograph Working Group. "Carcinogenicity of Consumption of Red and Processed Meat." *Lancet Oncology* 16, no. 16 (December 2015): 1599–1600.

"BRCA Mutations: Cancer Risk and Genetic Testing." *National Cancer Institute*, February 5, 2018. https://www.cancer.gov/about-cancer/causes-prevention/genetics/brca-fact-sheet.

Brown, J. S., and C. A. Aktipis. "Inclusive Fitness Effects Can Select for Cancer Suppression into Old Age." *Philosophical Transactions of the Royal Society of London, Series B: Biological Sciences* 370, no. 1673 (2015). https://doi.org/10.1098/rstb.2015.0160.

Broxterman, H. J., H. M. Pinedo, C. M. Kuiper, L. C. Kaptein, G. J. Schuurhuis, and J. Lankelma. "Induction by Verapamil of a Rapid Increase in ATP Consumption in Multidrug-Resistant Tumor Cells." *FASEB Journal* 2, no. 7 (April 1988): 2278–82.

Bullock, A. N., and A. R. Fersht. "Rescuing the Function of Mutant p53." *Nature Reviews Cancer* 1, no. 1 (October 2001): 68–76.

Burns, Kathleen H. "Transposable Elements in Cancer." *Nature Reviews Cancer* 17, no. 7 (July 2017): 415–24.

Buss, L. W. "Somatic Cell Parasitism and the Evolution of Somatic Tissue Compatibility." *Proceedings of the National Academy of Sciences of the United States of America* 79, no. 17 (September 1982): 5337–41.

Buss, Leo W. *The Evolution of Individuality*. Princeton, NJ: Princeton University Press, 1987.

Cairns, J. "Mutation Selection and the Natural History of Cancer." *Nature* 255, no. 5505 (1975): 197–200.

Campisi, Judith. "Cancer and Ageing: Rival Demons?" *Nature Reviews Cancer* 3, no. 5 (May 2003): 339–49.

Capasso, Luigi L. "Antiquity of Cancer." *International Journal of Cancer* 113, no. 1 (January 2005): 2–13.

"Carcinogens Listed in the Eleventh Report." In *The Report on Carcinogens*, 11th ed. Durham, NC: National Toxicology Program, U.S. Department of Health and Human Services, 2011. https://web.archive.org/web/20090507123840if_/http://ntp.niehs.nih.gov/ntp/roc/eleventh/known.pdf.

Caulin, A. F., and C. C. Maley. "Peto's Paradox: Evolution's Prescription for Cancer Prevention." *Trends in Ecology and Evolution* 26, no. 4 (February 2011): 175–82.

Cavallo, F., C. De Giovanni, P. Nanni, G. Forni, and P. L. Lollini. "2011: The Immune Hallmarks of Cancer." *Cancer Immunology, Immunotherapy* 60, no. 3 (2011): 319–26.

Chakrabarty, A. M. "Microorganisms and Cancer: Quest for a Therapy." *Journal of Bacteriology* 185, no. 9 (May 2003): 2683–86.

Chapman, Anna, Laura Fernandez del Ama, Jennifer Ferguson, Jivko Kamarashev, Claudia Wellbrock, and Adam Hurlstone. "Heterogeneous Tumor Subpopulations Cooperate to Drive Invasion." *Cell Reports* 8, no. 3 (August 2014): 688–95.

Chiarella, Paula, Juan Bruzzo, Roberto P. Meiss, and Raúl A. Ruggiero. "Concomitant Tumor Resistance." *Cancer Letters* 324, no. 2 (November 2012): 133–41.

Churchill, Winston S. November 11, 1947. The International Churchill Society. https://winstonchurchill.org/resources/quotes/the-worst-form-of-government/.

Clavel-Chapelon, F., and E3N Group. "Cumulative Number of Menstrual Cycles and Breast Cancer Risk: Results from the E3N Cohort Study of French Women." *Cancer Causes and Control* 13, no. 9 (November 2002): 831–38.

Cline, Melissa S., Rachel G. Liao, Michael T. Parsons, Benedict Paten, Faisal Alquaddoomi, Antonis Antoniou, Samantha Baxter, et al. "BRCA Challenge: BRCA Exchange as a Global Resource for Variants in BRCA1 and BRCA2." *PLoS Genetics* 14, no. 12 (December 2018): e1007752.

Coan, P. M., G. J. Burton, and A. C. Ferguson-Smith. "Imprinted Genes in the Placenta—A Review." *Placenta* 26, suppl. A (2005): S10–S20.

Collado, Manuel, Maria A. Blasco, and Manuel Serrano. "Cellular Senescence in Cancer and Aging." *Cell* 130, no. 2 (July 2007): 223–33.

Connor, Stephen R., Bruce Pyenson, Kathryn Fitch, Carol Spence, and Kosuke Iwasaki. "Comparing Hospice and Nonhospice Patient Survival among Patients Who Die within a Three-Year Window." *Journal of Pain and Symptom Management* 33, no. 3 (March 2007): 238–46.

Correa, Raul, Philip C. Thornton, Susan M. Rosenberg, and P. J. Hastings. "Oxygen and RNA in Stress-Induced Mutation." *Current Genetics* 64, no. 4 (August 2018): 769–76.

Cougnoux, Antony, Guillaume Dalmasso, Ruben Martinez, Emmanuel Buc, Julien Delmas, Lucie Gibold, Pierre Sauvanet, et al. "Bacterial Genotoxin Colibactin Promotes Colon Tumour Growth by Inducing a Senescence-Associated Secretory Phenotype." *Gut* 63, no. 12 (December 2014): 1932–42.

Cronk, Lee, Colette Berbesque, Thomas Conte, Matthew Gervais, Padmini Iyer, Brighid McCarthy, Dennis Sonkoi, Cathryn Townsend, and Athena Aktipis. "Managing Risk through Cooperation: Need-Based Transfers and Risk Pooling among the Societies of the Human Generosity Project." In *Global Perspectives on Long-Term Community Resource Management*, edited by L. Lozny and T. McGovern, 41–75. New York: Springer, 2019.

Cuzick, Jack, Florian Otto, John A. Baron, Powel H. Brown, John Burn, Peter Greenwald, Janusz Jankowski, et al. "Aspirin and Non-Steroidal Anti-Inflammatory Drugs for Cancer Prevention: An International Consensus Statement." *Lancet Oncology* 10, no. 5 (May 2009): 501–7.

Dalmasso, Guillaume, Antony Cougnoux, Julien Delmas, Arlette Darfeuille-Michaud, and Richard Bonnet. "The Bacterial Genotoxin Colibactin Promotes Colon Tumor Growth by Modifying the Tumor Microenvironment." *Gut Microbes* 5, no. 5 (2014): 675–80.

Daum, Hagit, Tamar Peretz, and Neri Laufer. "BRCA Mutations and Reproduction." *Fertility and Sterility* 109, no. 1 (January 2018): 33–38.

Davis, S., D. K. Mirick, and R. G. Stevens. "Night Shift Work, Light at Night, and Risk of Breast Cancer." *Journal of the National Cancer Institute* 93, no. 20 (October 2001): 1557–62.

Dawkins, Richard. *The Selfish Gene*. Oxford: Oxford University Press, 1976.

de Martel, Catherine, Jacques Ferlay, Silvia Franceschi, Jérôme Vignat, Freddie Bray, David Forman, and Martyn Plummer. "Global Burden of Cancers Attributable to Infections in 2008: A Review and Synthetic Analysis." *Lancet Oncology* 13, no. 6 (June 2012): 607–15.

Diaz, Luis A., Jr., Richard T. Williams, Jian Wu, Isaac Kinde, J. Randolph Hecht, Jordan Berlin, Benjamin Allen, et al. "The Molecular Evolution of Acquired Resistance to Targeted EGFR Blockade in Colorectal Cancers." *Nature* 486, no. 7404 (June 2012): 537–40.

Dobzhansky, Theodosius. "Nothing in Biology Makes Sense Except in the Light of Evolution." *American Biology Teacher* 35, no. 3 (March 1973): 125–29.

Dugatkin, Lee Alan. "Animal Cooperation among Unrelated Individuals." *Die Naturwissenschaften* 89, no. 12 (December 2002): 533–41.

Dunn, Gavin P., Allen T. Bruce, Hiroaki Ikeda, Lloyd J. Old, and Robert D. Schreiber. "Cancer Immunoediting: From Immunosurveillance to Tumor Escape." *Nature Immunology* 3, no. 11 (November 2002): 991–98.

Dvorak, Harold F. "Tumors: Wounds That Do Not Heal." *New England Journal of Medicine* 315, no. 26 (December 1986): 1650–59.

Eisenhoffer, George T., Patrick D. Loftus, Masaaki Yoshigi, Hideo Otsuna, Chi-Bin Chien, Paul A. Morcos, and Jody Rosenblatt. "Crowding Induces Live Cell Extrusion to Maintain Homeostatic Cell Numbers in Epithelia." *Nature* 484, no. 7395 (April 2012): 546–49.

Elser, James J., Marcia M. Kyle, Marilyn S. Smith, and John D. Nagy. "Biological Stoichiometry in Human Cancer." *PloS One* 2, no. 10 (2007): e1028.

Ember, Melvin, Carol R. Ember, and Bobbi S. Low. "Comparing Explanations of Polygyny." *Cross-Cultural Research* 41, no. 4 (November 2007): 428–40.

Enriquez-Navas, Pedro M., Yoonseok Kam, Tuhin Das, Sabrina Hassan, Ariosto Silva, Parastou Foroutan, Epifanio Ruiz, et al. "Exploiting Evolutionary Principles to Prolong Tumor Control in Preclinical Models of Breast Cancer." *Science Translational Medicine* 8, no. 327 (February 2016): 327ra24.

Ewald, Paul W. "An Evolutionary Perspective on Parasitism as a Cause of Cancer." In *Advances in Parasitology*, vol. 68, 21–43. Cambridge, MA: Academic Press, 2009.

Featherston, J., and P. M. Durand. "Cooperation and Conflict in Cancer: An Evolutionary Perspective." *South African Journal of Science* 108, no. 9–10 (January 2012).

Fernandez, André A., and Paul R. Bowser. "Selection for a Dominant Oncogene and Large Male Size as a Risk Factor for Melanoma in the Xiphophorus Animal Model." *Molecular Ecology* 19, no. 15 (August 2010): 3114–23.

Ferreri, Andrés J. M., Silvia Govi, Elisa Pasini, Silvia Mappa, Francesco Bertoni, Francesco Zaja, Carlos Montalbán, et al. "Chlamydophila Psittaci Eradication with Doxycycline as First-Line Targeted Therapy for Ocular Adnexae Lymphoma: Final Results of an International Phase II Trial." *Journal of Clinical Oncology* 30, no. 24 (August 2012): 2988–94.

Finn, Olivera J. "Human Tumor Antigens Yesterday, Today, and Tomorrow." *Cancer Immunology Research* 5, no. 5 (May 2017): 347–54.

Fleming, J. M., K. E. Creevy, and D. E. L. Promislow. "Mortality in North American Dogs from 1984 to 2004: An Investigation into Age-, Size-, and Breed-Related Causes of Death." *Journal of Veterinary Internal Medicine* 25, no. 2 (March 2011): 187–98.

Fletcher, J. A., and Michael Doebeli. "A Simple and General Explanation for the Evolution of Altruism." *Proceedings of the Royal Society B: Biological Sciences* 276, no. 1654 (2009): 13–19.

Flossmann, Enrico, Peter M. Rothwell, and British Doctors Aspirin Trial and the U.K.-TIA Aspirin Trial. "Effect of Aspirin on Long-Term Risk of Colorectal Cancer: Consistent Evidence from Randomised and Observational Studies." *Lancet* 369, no. 9573 (May 2007): 1603–13.

Frank, S. A., and M. A. Nowak. "Cell Biology: Developmental Predisposition to Cancer." *Nature* 422, no. 6931 (2003): 494.

Gabrilovich, D., and V. Pisarev. "Tumor Escape from Immune Response: Mechanisms and Targets of Activity." *Current Drug Targets* 4, no. 7 (2003): 525–36.

Gaines, Sara, Ashley J. Williamson, Neil Hyman, and Jessica Kandel. "How the Microbiome Is Shaping Our Understanding of Cancer Biology and Its Treatment." *Seminars in Colon and Rectal Surgery* 29, no. 1 (March 2018): 12–16.

Gale, K. B., A. M. Ford, R. Repp, A. Borkhardt, C. Keller, O. B. Eden, and M. F. Greaves. "Backtracking Leukemia to Birth: Identification of Clonotypic Gene Fusion Sequences in Neonatal Blood Spots." *Proceedings of the National Academy of Sciences of the United States of America* 94, no. 25 (December 1997): 13950–54.

Gandhi, Manish J., and D. Michael Strong. "Donor Derived Malignancy Following Transplantation: A Review." *Cell and Tissue Banking* 8, no. 4 (April 2007): 267–86.

Gärtner, H. V., C. Seidl, C. Luckenbach, G. Schumm, E. Seifried, H. Ritter, and B. Bültmann. "Genetic Analysis of a Sarcoma Accidentally Transplanted from a

Patient to a Surgeon." *New England Journal of Medicine* 335, no. 20 (November 1996): 1494–96.

Gatenby, Robert A. "A Change of Strategy in the War on Cancer." *Nature* 459, no. 7246 (2009): 508–9.

Gatenby, Robert A., and Robert J. Gillies. "A Microenvironmental Model of Carcinogenesis." *Nature Reviews Cancer* 8, no. 1 (January 2008): 56–61.

Gatenby, Robert A., Ariosto S. Silva, Robert J. Gillies, and B. Roy Frieden. "Adaptive Therapy." *Cancer Research* 69, no. 11 (June 2009): 4894–903.

Gerlinger, M., A. J. Rowan, S. Horswell, J. Larkin, D. Endesfelder, E. Gronroos, P. Martinez, et al. "Intratumor Heterogeneity and Branched Evolution Revealed by Multiregion Sequencing." *New England Journal of Medicine* 366, no. 10 (2012): 883–92.

Gerlinger, Marco, Stuart Horswell, James Larkin, Andrew J. Rowan, Max P. Salm, Ignacio Varela, Rosalie Fisher, et al. "Genomic Architecture and Evolution of Clear Cell Renal Cell Carcinomas Defined by Multiregion Sequencing." *Nature Genetics* 46, no. 3 (March 2014): 225–33.

Germain, Pierre-Luc, and Lucie Laplane. "Metastasis as Supra-Cellular Selection? A Reply to Lean and Plutynski." *Biology and Philosophy* 32, no. 2 (March 2017): 281–87.

Ghoul, Melanie, Ashleigh S. Griffin, and Stuart A. West. "Toward an Evolutionary Definition of Cheating." *Evolution: International Journal of Organic Evolution* 68, no. 2 (February 2014): 318–31.

Goodman, Brian, and Humphrey Gardner. "The Microbiome and Cancer." *Journal of Pathology* 244, no. 5 (April 2018): 667–76.

Goodwin, Andrew C., Christina E. Destefano Shields, Shaoguang Wu, David L. Huso, Xinqun Wu, Tracy R. Murray-Stewart, Amy Hacker-Prietz, et al. "Polyamine Catabolism Contributes to Enterotoxigenic Bacteroides Fragilis-Induced Colon Tumorigenesis." *Proceedings of the National Academy of Sciences of the United States of America* 108, no. 37 (September 2011): 15354–59.

Gould, Stephen J. "The Median Isn't the Message." *Discover* 6, no. 6 (1985): 40–42.

Greaves, M. "Does Everyone Develop Covert Cancer?" *Nature Reviews Cancer* 14, no. 4 (2014): 209–10.

Greaves, M., and C. C. Maley. "Clonal Evolution in Cancer." *Nature* 481 (2012): 306–13.

Greaves, M. F. *Cancer: The Evolutionary Legacy*. Oxford: Oxford University Press, 2000.

Greaves, Mel. "A Causal Mechanism for Childhood Acute Lymphoblastic Leukaemia." *Nature Reviews Cancer* 18, no. 8 (August 2018): 471–84.

Greaves, Mel, and William Hughes. "Cancer Cell Transmission via the Placenta." *Evolution, Medicine, and Public Health* 2018, no. 1 (April 2018): 106–15.

Greaves, Mel F., Ana Teresa Maia, Joseph L. Wiemels, and Anthony M. Ford. "Leukemia in Twins: Lessons in Natural History." *Blood* 102, no. 7 (October 2003): 2321–33.

Green, Jane, Benjamin J. Cairns, Delphine Casabonne, F. Lucy Wright, Gillian Reeves, and Valerie Beral. "Height and Cancer Incidence in the Million Women

Study: Prospective Cohort, and Meta-Analysis of Prospective Studies of Height and Total Cancer Risk." *Lancet Oncology* 12, no. 8 (August 2011): 785–94.

Grosberg, Richard K., and Richard R. Strathmann. "The Evolution of Multicellularity: A Minor Major Transition?" *Annual Review of Ecology, Evolution, and Systematics* 38, no. 1 (December 2007): 621–54.

Gugel, E. A., and M. E. Sanders. "Needle-Stick Transmission of Human Colonic Adenocarcinoma." *New England Journal of Medicine* 315, no. 23 (December 1986): 1487.

Gurven, M., and H. Kaplan. "Longevity among Hunter-Gatherers: A Cross-Cultural Examination." *Population and Development Review* 33, no. 2 (2007): 321–65. https://onlinelibrary.wiley.com/doi/abs/10.1111/j.1728-4457.2007.00171.x.

Haeger, Anna, Katarina Wolf, Mirjam M. Zegers, and Peter Friedl. "Collective Cell Migration: Guidance Principles and Hierarchies." *Trends in Cell Biology* 25, no. 9 (September 2015): 556–66.

Haig, David. "Genomic Imprinting and the Theory of Parent-Offspring Conflict." *Seminars in Developmental Biology* 3 (1992): 153–60.

Haig, David. "Maternal-Fetal Conflict, Genomic Imprinting and Mammalian Vulnerabilities to Cancer." *Philosophical Transactions of the Royal Society of London, Series B: Biological Sciences* 370, no. 1673 (July 2015). https://doi.org/10.1098/rstb.2014.0178.

Hamilton, W. D. "The Genetical Evolution of Social Behaviour. I." *Journal of Theoretical Biology* 7, no. 1 (July 1964): 1–16.

Hamilton, W. D. "The Genetical Evolution of Social Behaviour. II." *Journal of Theoretical Biology* 7, no. 1 (July 1964): 1–16.

Hamilton, W. D., R. Axelrod, and R. Tanese. "Sexual Reproduction as an Adaptation to Resist Parasites (a Review)." *Proceedings of the National Academy of Sciences of the United States of America* 87, no. 9 (May 1990): 3566–73.

Hanahan, D., and R. A. Weinberg. "The Hallmarks of Cancer." *Cell* 100, no. 1 (2000): 57–70.

Hanahan, Douglas, and Robert A. Weinberg. "Hallmarks of Cancer: The Next Generation." *Cell* 144, no. 5 (March 2011): 646–74.

Hansen, Elsa, Robert J. Woods, and Andrew F. Read. "How to Use a Chemotherapeutic Agent When Resistance to It Threatens the Patient." *PLoS Biology* 15, no. 2 (2017): e2001110.

Hardy, Kathy, and Philip John Hardy. "1st Trimester Miscarriage: Four Decades of Study." *Translational Pediatrics* 4, no. 2 (April 2015): 189–200.

Hauser, David J., and Norbert Schwarz. "The War on Prevention: Bellicose Cancer Metaphors Hurt (Some) Prevention Intentions." *Personality and Social Psychology Bulletin* 41, no. 1 (January 2015): 66–77.

Hawkes, K., J. F. O'Connell, N. G. Jones, H. Alvarez, and E. L. Charnov. "Grandmothering, Menopause, and the Evolution of Human Life Histories." *Proceedings of the National Academy of Sciences of the United States of America* 95, no. 3 (February 1998): 1336–39.

Health Resources and Services Administration. "Organ Donation Statistics." Accessed December 19, 2017. https://www.organdonor.gov/statistics-stories /statistics.html.

Higa, Kelly C., and James DeGregori. "Decoy Fitness Peaks, Tumor Suppression, and Aging." *Aging Cell* 18, no. 3 (June 2019): e12938.

Hodi, F. Stephen, Steven J. O'Day, David F. McDermott, Robert W. Weber, Jeffrey A. Sosman, John B. Haanen, Rene Gonzalez, et al. "Improved Survival with Ipilimumab in Patients with Metastatic Melanoma." *New England Journal of Medicine* 363, no. 8 (August 2010): 711–23.

Hölldobler, Bert, and Edward O. Wilson. *The Superorganism: The Beauty, Elegance, and Strangeness of Insect Societies*. New York: W. W. Norton and Company, 2009.

Hosseini, Hedayatollah, Milan M. S. Obradović, Martin Hoffmann, Kathryn L. Harper, Maria Soledad Sosa, Melanie Werner-Klein, Lahiri Kanth Nanduri, et al. "Early Dissemination Seeds Metastasis in Breast Cancer." *Nature* 540, no. 7634 (December 2016): 552–58.

Huang, Sui. "The War on Cancer: Lessons from the War on Terror." *Frontiers in Oncology* 4 (October 2014): 293.

Huang, Yuhui, Shom Goel, Dan G. Duda, Dai Fukumura, and Rakesh K. Jain. "Vascular Normalization as an Emerging Strategy to Enhance Cancer Immunotherapy." *Cancer Research* 73, no. 10 (May 2013): 2943–48.

Hussell, T., P. G. Isaacson, J. E. Crabtree, and J. Spencer. "Helicobacter Pylori-Specific Tumour-Infiltrating T Cells Provide Contact Dependent Help for the Growth of Malignant B Cells in Low-Grade Gastric Lymphoma of Mucosa-Associated Lymphoid Tissue." *Journal of Pathology* 178, no. 2 (February 1996): 122–27.

Iida, Noriho, Amiran Dzutsev, C. Andrew Stewart, Loretta Smith, Nicolas Bouladoux, Rebecca A. Weingarten, Daniel A. Molina, et al. "Commensal Bacteria Control Cancer Response to Therapy by Modulating the Tumor Microenvironment." *Science* 342, no. 6161 (November 2013): 967–70.

Jansen, G., R. Gatenby, and C. A. Aktipis. "Opinion: Control vs. Eradication: Applying Infectious Disease Treatment Strategies to Cancer." *Proceedings of the National Academy of Sciences of the United States of America* 112, no. 4 (2015): 937–38.

Jenkins, Russell W., David A. Barbie, and Keith T. Flaherty. "Mechanisms of Resistance to Immune Checkpoint Inhibitors." *British Journal of Cancer* 118, no. 1 (January 2018): 9–16.

Johnson, A., and J. R. Giles. "The Hen as a Model of Ovarian Cancer." *Nature Reviews Cancer* 13, no. 6 (2013): 432–36.

Jones, Aria. "An Open Letter to People Who Use the 'Battle' Metaphor for Other People Who Have the Distinct Displeasure of Cancer." *McSweeney's Internet Tendency*. San Francisco: McSweeney's Publishing, October 19, 2012. https:// www.mcsweeneys.net/articles/an-open-letter-to-people-who-use-the-battle -metaphor-for-other-people-who-have-the-distinct-displeasure-of-cancer.

Jones, S., W. D. Chen, G. Parmigiani, F. Diehl, N. Beerenwinkel, T. Antal, A. Traulsen, et al. "Comparative Lesion Sequencing Provides Insights into Tumor Evolution." *Proceedings of the National Academy of Sciences of the United States of America* 105, no. 11 (2008): 4283–88.

Kajita, Mihoko, Kaoru Sugimura, Atsuko Ohoka, Jemima Burden, Hitomi Suganuma, Masaya Ikegawa, Takashi Shimada, et al. "Filamin Acts as a Key Regulator in Epithelial Defence against Transformed Cells." *Nature Communications* 5 (July 2014): 4428.

Kam, Yoonseok, Tuhin Das, Haibin Tian, Parastou Foroutan, Epifanio Ruiz, Gary Martinez, Susan Minton, Robert J. Gillies, and Robert A. Gatenby. "Sweat but No Gain: Inhibiting Proliferation of Multidrug Resistant Cancer Cells with 'ersatzdroges.'" *International Journal of Cancer* 136, no. 4 (2015): E188–96.

Kauffman, H. Myron, Maureen A. McBride, Wida S. Cherikh, Pamela C. Spain, and Francis L. Delmonico. "Transplant Tumor Registry: Donors with Central Nervous System Tumors." *Transplantation* 73, no. 4 (February 2002): 579–82.

Kauffman, H. Myron, Maureen A. McBride, Wida S. Cherikh, Pamela C. Spain, William H. Marks, and Allan M. Roza. "Transplant Tumor Registry: Donor Related Malignancies." *Transplantation* 74, no. 3 (August 2002): 358–62.

Keoun, Brad. "Ashkenazim Not Alone: Other Ethnic Groups Have Breast Cancer Gene Mutations, Too." *Journal of the National Cancer Institute* 89, no. 1 (January 1997): 8–9.

Knoll, Andrew H., and David Hewitt. "Phylogenetic, Functional and Geological Perspectives on Complex Multicellularity." In *The Major Transitions in Evolution Revisited*, edited by Brett Calcott and Kim Sterelny, 251–70. Cambridge, MA: MIT Press, 2011.

Kostadinov, Rumen L., Mary K. Kuhner, Xiaohong Li, Carissa A. Sanchez, Patricia C. Galipeau, Thomas G. Paulson, Cassandra L. Sather, et al. "NSAIDs Modulate Clonal Evolution in Barrett's Esophagus." *PLoS Genetics* 9, no. 6 (June 2013): e1003553.

Kuchenbaecker, Karoline B., John L. Hopper, Daniel R. Barnes, Kelly-Anne Phillips, Thea M. Mooij, Marie-José Roos-Blom, Sarah Jervis, et al. "Risks of Breast, Ovarian, and Contralateral Breast Cancer for BRCA1 and BRCA2 Mutation Carriers." *Journal of the American Medical Association* 317, no. 23 (June 2017): 2402–16.

Kurbel, Sven, Stjepko Plestina, and Damir Vrbanec. "Occurrence of the Acquired Immunity in Early Vertebrates due to Danger of Transmissible Cancers Similar to Canine Venereal Tumors." *Medical Hypotheses* 68, no. 5 (2007): 1185–86.

Kurian, Allison W., Yun Li, Ann S. Hamilton, Kevin C. Ward, Sarah T. Hawley, Monica Morrow, M. Chandler McLeod, Reshma Jagsi, and Steven J. Katz. "Gaps in Incorporating Germline Genetic Testing into Treatment Decision-Making for Early-Stage Breast Cancer." *Journal of Clinical Oncology* 35, no. 20 (July 2017): 2232–39.

Kwiatkowski, Fabrice, Marie Arbre, Yannick Bidet, Claire Laquet, Nancy Uhrhammer, and Yves-Jean Bignon. "BRCA Mutations Increase Fertility in Families at Hereditary Breast/Ovarian Cancer Risk." *PloS One* 10, no. 6 (June 2015): e0127363.

Landis, T. D., and R. K. Dumroese, "Integrated Pest Management—An Overview and Update." *Forest Nursery Notes* (2014). https://www.researchgate.net/profile /R_Kasten_Dumroese/publication/272682105_Integrated_pest_management -an_overview_and_update/links/54ebbce10cf2082851be7e2b.pdf.

Lean, Christopher, and Anya Plutynski. "The Evolution of Failure: Explaining Cancer as an Evolutionary Process." *Biology and Philosophy* 31, no. 1 (January 2016): 39–57.

Lecuit, Marc, Eric Abachin, Antoine Martin, Claire Poyart, Philippe Pochart, Felipe Suarez, Djaouida Bengoufa, et al. "Immunoproliferative Small Intestinal Disease Associated with Campylobacter Jejuni." *New England Journal of Medicine* 350, no. 3 (January 2004): 239–48.

"Leukemia—Cancer Stat Facts." Surveillance, Epidemiology, and End Results Program, National Cancer Institute, accessed June 20, 2019, https://seer.cancer .gov/statfacts/html/leuks.html.

Luebeck, Georg E., William D. Hazelton, Kit Curtius, Sean K. Maden, Ming Yu, Kelly T. Carter, Wynn Burke, et al. "Implications of Epigenetic Drift in Colorectal Neoplasia." *Cancer Research* 79, no. 3 (February 2019): 495–504.

Maley, Carlo C., Athena Aktipis, Trevor A. Graham, Andrea Sottoriva, Amy M. Boddy, Michalina Janiszewska, Ariosto S. Silva, et al. "Classifying the Evolutionary and Ecological Features of Neoplasms." *Nature Reviews Cancer* 17, no. 10 (October 2017): 605–19.

Marin, Ioana, and Jonathan Kipnis. "Learning and Memory . . . and the Immune System." *Learning and Memory* 20, no. 10 (September 2013): 601–6.

Marlowe, Frank W. "The Mating System of Foragers in the Standard Cross-Cultural Sample." *Cross-Cultural Research* 37, no. 3 (August 2003): 282–306.

Martincorena, Inigo, Joanna C. Fowler, Agnieszka Wabik, Andrew R. J. Lawson, Federico Abascal, Michael W. J. Hall, Alex Cagan, et al. "Somatic Mutant Clones Colonize the Human Esophagus with Age." *Science* 362, no. 6417 (November 2018): 911–17.

Martincorena, Inigo, Amit Roshan, Moritz Gerstung, Peter Ellis, Peter Van Loo, Stuart McLaren, David C. Wedge, et al. "Tumor Evolution: High Burden and Pervasive Positive Selection of Somatic Mutations in Normal Human Skin." *Science* 348, no. 6237 (May 2015): 880–86.

Marusyk, Andriy, Doris P. Tabassum, Philipp M. Altrock, Vanessa Almendro, Franziska Michor, and Kornelia Polyak. "Non-Cell-Autonomous Driving of Tumour Growth Supports Sub-Clonal Heterogeneity." *Nature* 514, no. 7520 (October 2014): 54–58.

Maynard Smith, John. "Group Selection and Kin Selection." *Nature* 201 (March 1964): 1145.

Maynard Smith, John, and Eörs Szathmáry. *The Major Transitions in Evolution.* Oxford: Oxford University Press, 1995.

Mazzone, M., D. Dettori, R. Leite de Oliveira, S. Loges, T. Schmidt, B. Jonckx, Y. M. Tian, et al. "Heterozygous Deficiency of PHD2 Restores Tumor Oxygenation and Inhibits Metastasis via Endothelial Normalization." *Cell* 136, no. 5 (2009): 839–51.

McKeown, Thomas, and R. G. Record. "The Influence of Placental Size on Foetal Growth according to Sex and Order of Birth." *Journal of Endocrinology* 10, no. 1 (November 1953): 73–81.

Meier-Abt, Fabienne, Mohamed Bentires-Alj, and Christoph Rochlitz. "Breast Cancer Prevention: Lessons to Be Learned from Mechanisms of Early Pregnancy-Mediated Breast Cancer Protection." *Cancer Research* 75, no. 5 (March 2015): 803–7.

Merlo, Lauren F., J. W. Pepper, Brian J. Reid, and Carlo C. Maley. "Cancer as an Evolutionary and Ecological Process." *Nature Reviews Cancer* 6, no. 12 (2006): 924–35.

Metzger, Michael J., and Stephen P. Goff. "A Sixth Modality of Infectious Disease: Contagious Cancer from Devils to Clams and Beyond." *PLoS Pathogens* 12, no. 10 (October 2016): e1005904.

Meza, R., E. G. Luebeck, and S. H. Moolgavkar. "Gestational Mutations and Carcinogenesis." *Mathematical Biosciences* 197, no. 2 (2005): 188–210.

Moller, Henrik. "Lessons for Invasion Theory from Social Insects." *Biological Conservation* 78, no. 1 (October 1996): 125–42.

Monk, M., and C. Holding. "Human Embryonic Genes Re-Expressed in Cancer Cells." *Oncogene* 20, no. 56 (December 2001): 8085–91.

Moore, A. E., C. P. Rhoads, and C. M. Southam. "Homotransplantation of Human Cell Lines." *Science* 125, no. 3239 (January 1957): 158–60.

Morange, Michel. "What History Tells Us XXVIII. What Is Really New in the Current Evolutionary Theory of Cancer?" *Journal of Biosciences* 37, no. 4 (September 2012): 609–12.

Moslehi, Roxana, Ranjana Singh, Lawrence Lessner, and Jan M. Friedman. "Impact of BRCA Mutations on Female Fertility and Offspring Sex Ratio." *American Journal of Human Biology* 22, no. 2 (March 2010): 201–5.

Muehlenbachs, Atis, Julu Bhatnagar, Carlos A. Agudelo, Alicia Hidron, Mark L. Eberhard, Blaine A. Mathison, Michael A. Frace, et al. "Malignant Transformation of Hymenolepis Nana in a Human Host." *New England Journal of Medicine* 373, no. 19 (November 2015): 1845–52.

Muñoz, Nubia, Xavier Castellsagué, Amy Berrington de González, and Lutz Gissmann. "Chapter 1: HPV in the Etiology of Human Cancer." *Vaccine* 24, suppl. 3 (August 2006): S3/1–10.

Murchison, E. P. "Clonally Transmissible Cancers in Dogs and Tasmanian Devils." *Oncogene* 27, suppl. 2 (December 2008): S19–S30.

Murgia, Claudio, Jonathan K. Pritchard, Su Yeon Kim, Ariberto Fassati, and Robin A. Weiss. "Clonal Origin and Evolution of a Transmissible Cancer." *Cell* 126, no. 3 (August 2006): 477–87.

Nahta, Rita, Dihua Yu, Mien-Chie Hung, Gabriel N. Hortobagyi, and Francisco J. Esteva. "Mechanisms of Disease: Understanding Resistance to HER2-Targeted Therapy in Human Breast Cancer." *Nature Clinical Practice Oncology* 3, no. 5 (May 2006): 269–80.

National Cancer Institute. "NCI Dictionary of Cancer Terms." Accessed February 2, 2011. https://www.cancer.gov/publications/dictionaries/cancer-terms.

Nesse, Randolph M. "Natural Selection and the Regulation of Defenses: A Signal Detection Analysis of the Smoke Detector Principle." *Evolution and Human Behavior* 26, no. 1 (2005): 88–105.

Noë, Ronald, and Peter Hammerstein. "Biological Markets: Supply and Demand Determine the Effect of Partner Choice in Cooperation, Mutualism and Mating." *Behavioral Ecology and Sociobiology* 35, no. 1 (1994): 1–11.

Nougayrède, Jean-Philippe, Stefan Homburg, Frédéric Taieb, Michèle Boury, Elzbieta Brzuszkiewicz, Gerhard Gottschalk, Carmen Buchrieser, Jörg Hacker, Ulrich Dobrindt, and Eric Oswald. "Escherichia Coli Induces DNA Double-Strand Breaks in Eukaryotic Cells." *Science* 313, no. 5788 (August 2006): 848–51.

Nowell, Peter C. "The Clonal Evolution of Tumor Cell Populations." *Science* 194, no. 4260 (1976): 23–28.

Nunney, Leonard. "Lineage Selection and the Evolution of Multistage Carcinogenesis." *Proceedings of the Royal Society of London, Series B* 266, no. 1418 (March 7, 1999): 493–98.

Nunney, Leonard, Carlo C. Maley, Matthew Breen, Michael E. Hochberg, and Joshua D. Schiffman. "Peto's Paradox and the Promise of Comparative Oncology." *Philosophical Transactions of the Royal Society of London, Series B: Biological Sciences* 370, no. 1673 (July 2015). https://doi.org/10.1098/rstb.2014.0177.

Odes, Edward J., Patrick S. Randolph-Quinney, Maryna Steyn, Zach Throckmorton, Jacqueline S. Smilg, Bernhard Zipfel, Tanya N. Augustine, Frikkie de Beer, et al. "Earliest Hominin Cancer: 1.7-Million-Year-Old Osteosarcoma from Swartkrans Cave, South Africa." *South African Journal of Science* 112, no. 7/8 (July 2016). https://doi.org/10.17159/sajs.2016/20150471.

Office for National Statistics. "Causes of Death over 100 Years." September 18, 2017. https://www.ons.gov.uk/peoplepopulationandcommunity/births deathsandmarriages/deaths/articles/causesofdeathover100years/2017-09-18.

Olson, Peter D., Kristine Yoder, Luis F. Fajardo, Aileen M. Marty, Simone van de Pas, Claudia Olivier, and David A. Relman. "Lethal Invasive Cestodiasis in Immunosuppressed Patients." *Journal of Infectious Diseases* 187, no. 12 (June 2003): 1962–66.

Oronsky, Bryan, Corey A. Carter, Vernon Mackie, Jan Scicinski, Arnold Oronsky, Neil Oronsky, Scott Caroen, Christopher Parker, Michelle Lybeck, and Tony Reid. "The War on Cancer: A Military Perspective." *Frontiers in Oncology* 4 (2014): 387.

Pal, Tuya, David Keefe, Ping Sun, Steven A. Narod, and Hereditary Breast Cancer Clinical Study Group. "Fertility in Women with BRCA Mutations: A Case-Control Study." *Fertility and Sterility* 93, no. 6 (April 2010): 1805–8.

Pardoll, Drew M. "The Blockade of Immune Checkpoints in Cancer Immunotherapy." *Nature Reviews Cancer* 12, no. 4 (March 2012): 252–64.

Parvinen, Kalle. "Evolutionary Suicide." *Acta Biotheoretica* 53, no. 3 (2005): 241–64.

Penn, I., C. G. Halgrimson, and T. E. Starzl. "De Novo Malignant Tumors in Organ Transplant Recipients." *Transplantation Proceedings* 3, no. 1 (March 1971): 773–78.

Pepper, John W. "Drugs That Target Pathogen Public Goods Are Robust against Evolved Drug Resistance." *Evolutionary Applications* 5, no. 7 (November 2012): 757–61.

Perri, Angela, Chris Widga, Dennis Lawler, Terrance Martin, Thomas Loebel, Kenneth Farnsworth, Luci Kohn, and Brent Buenger. "New Evidence of the Earliest Domestic Dogs in the Americas." *bioRxiv*, June 27, 2018. https://doi.org/10.1101/343574.

Pesavento, Patricia A., Dalen Agnew, Michael K. Keel, and Kevin D. Woolard. "Cancer in Wildlife: Patterns of Emergence." *Nature Reviews Cancer* 18, no. 10 (October 2018): 646–61.

Peto, R., F. J. Roe, P. N. Lee, L. Levy, and J. Clack. "Cancer and Ageing in Mice and Men." *British Journal of Cancer* 32, no. 4 (October 1975): 411–26.

Peto, Richard. "Epidemiology, Multistage Models, and Short-Term Mutagenicity Tests." *International Journal of Epidemiology* 45, no. 3 (1977): 621–37.

Pfister, G. "Multisensor/Multicriteria Fire Detection: A New Trend Rapidly Becomes State of the Art." *Fire Technology* 33, no. 2 (May 1997): 115–39.

Pierce, Robert A., II, Jason Sumners, and Emily Flinn. "Antler Development in White-Tailed Deer: Implications for Management." University of Missouri Extension, January 2012. https://extension2.missouri.edu/g9486.

Proksch, Ehrhardt, Johanna M. Brandner, and Jens-Michael Jensen. "The Skin: An Indispensable Barrier." *Experimental Dermatology* 17, no. 12 (December 2008): 1063–72.

Pye, Ruth J., David Pemberton, Cesar Tovar, Jose M. C. Tubio, Karen A. Dun, Samantha Fox, Jocelyn Darby, et al. "A Second Transmissible Cancer in Tasmanian Devils." *Proceedings of the National Academy of Sciences of the United States of America* 113, no. 2 (January 2016): 374–79.

Quinlan, Robert J., and Marsha B. Quinlan. "Evolutionary Ecology of Human Pair-Bonds: Cross-Cultural Tests of Alternative Hypotheses." *Cross-Cultural Research* 41, no. 2 (May 2007): 149–69.

Rankin, Erinn B., and Amato J. Giaccia. "Hypoxic Control of Metastasis." *Science* 352, no. 6282 (April 2016): 175–80.

Read, Andrew F. "The Selfish Germ." *PLoS Biology* 15, no. 7 (July 2017): e2003250.

Rebbeck, Clare A., Rachael Thomas, Matthew Breen, Armand M. Leroi, and Austin Burt. "Origins and Evolution of a Transmissible Cancer." *Evolution: International Journal of Organic Evolution* 63, no. 9 (September 2009): 2340–49.

Reik, Wolf, Miguel Constância, Abigail Fowden, Neil Anderson, Wendy Dean, Anne Ferguson-Smith, Benjamin Tycko, and Colin Sibley. "Regulation of Supply and Demand for Maternal Nutrients in Mammals by Imprinted Genes." *Journal of Physiology* 547, pt. 1 (February 2003): 35–44.

Robey, Ian F., Brenda K. Baggett, Nathaniel D. Kirkpatrick, Denise J. Roe, Julie Dosescu, Bonnie F. Sloane, Arig Ibrahim Hashim, et al. "Bicarbonate Increases Tumor pH and Inhibits Spontaneous Metastases." *Cancer Research* 69, no. 6 (March 2009): 2260–68.

Rogozin, Igor B., and Youri I. Pavlov. "Theoretical Analysis of Mutation Hotspots and Their DNA Sequence Context Specificity." *Mutation Research* 544, no. 1 (September 2003): 65–85.

Rosalie, David A., and Michael R. Zimmerman. "Cancer: An Old Disease, a New Disease or Something in Between?" *Nature Reviews Cancer* 10, no. 10 (2010): 728–33.

Rosenberg, S. M. "Evolving Responsively: Adaptive Mutation." *Nature Reviews Genetics* 2, no. 7 (July 2001): 504–15.

Rothschild, Bruce M., Brian J. Witzke, and Israel Hershkovitz. "Metastatic Cancer in the Jurassic." *Lancet* 354, no. 9176 (July 1999): 398.

Rothwell, Peter M., F. Gerald R. Fowkes, Jill F. F. Belch, Hisao Ogawa, Charles P. Warlow, and Tom W. Meade. "Effect of Daily Aspirin on Long-Term Risk of Death due to Cancer: Analysis of Individual Patient Data from Randomised Trials." *Lancet* 377, no. 9759 (January 2011): 31–41.

Santamaría-Fríes, M., L. F. Fajardo, M. L. Sogin, P. D. Olson, and D. A. Relman. "Lethal Infection by a Previously Unrecognised Metazoan Parasite." *Lancet* 347, no. 9018 (June 1996): 1797–1801.

Scanlon, E. F., R. A. Hawkins, W. W. Fox, and W. S. Smith. "Fatal Homotransplanted Melanoma: A Case Report." *Cancer* 18 (June 1965): 782–89.

Schiffman, Joshua D., and Matthew Breen. "Comparative Oncology: What Dogs and Other Species Can Teach Us about Humans with Cancer." *Philosophical Transactions of the Royal Society of London, Series B: Biological Sciences* 370, no. 1673 (July 2015). https://doi.org/10.1098/rstb.2014.0231.

Schiffman, Joshua D., Richard M. White, Trevor A. Graham, Qihong Huang, and Athena Aktipis. "The Darwinian Dynamics of Motility and Metastasis." In *Frontiers in Cancer Research*, 135–76. New York: Springer, 2016.

Scott, Alasdair J., Claire A. Merrifield, Jessica A. Younes, and Elizabeth P. Pekelharing. "Pre-, Pro- and Synbiotics in Cancer Prevention and Treatment—A Review of Basic and Clinical Research." *ecancermedicalscience* 12 (September 2018): 869.

Siddle, Hannah V., and Jim Kaufman. "A Tale of Two Tumours: Comparison of the Immune Escape Strategies of Contagious Cancers." *Molecular Immunology* 55, no. 2 (September 2013): 190–93.

Siegel, R. L., K. D. Miller, and A. Jemal. "Cancer Statistics, 2018." *CA: A Cancer Journal for Clinicians* 68, no. 1 (2018): 7–30.

Smith, K. R., H. A. Hanson, G. P. Mineau, and S. S. Buys. "Effects of BRCA1 and BRCA2 Mutations on Female Fertility." *Proceedings of the Royal Society of London, Series B* 279, no. 1732 (2011): 1389–95. https://doi.org/10.1098/rspb .2011.1697.

Sober, Elliott, and David Sloan Wilson. *Unto Others: The Evolution and Psychology of Unselfish Behavior*. Cambridge, MA: Harvard University Press, 1998.

Sonnenschein, C., and A. M. Soto. *The Society of Cells: Cancer and Control of Cell Proliferation*. New York: Springer, 1999.

Sprouffske, Kathleen, C. Athena Aktipis, Jerald P. Radich, Martin Carroll, Aurora M. Nedelcu, and Carlo C. Maley. "An Evolutionary Explanation for the Presence of Cancer Nonstem Cells in Neoplasms." *Evolutionary Applications* 6, no. 1 (January 2013): 92–101.

Ståhl, Patrik L., Henrik Stranneheim, Anna Asplund, Lisa Berglund, Fredrik Pontén, and Joakim Lundeberg. "Sun-Induced Nonsynonymous p53 Mutations Are

Extensively Accumulated and Tolerated in Normal Appearing Human Skin." *Journal of Investigative Dermatology* 131, no. 2 (February 2011): 504–8.

Sulak, Michael, Lindsey Fong, Katelyn Mika, Sravanthi Chigurupati, Lisa Yon, Nigel P. Mongan, Richard D. Emes, and Vincent J. Lynch. "TP53 Copy Number Expansion Is Associated with the Evolution of Increased Body Size and an Enhanced DNA Damage Response in Elephants." *eLife* 5 (September 2016). https://doi.org/10.7554/eLife.11994.

Summers, K., J. da Silva, and M. A. Farwell. "Intragenomic Conflict and Cancer." *Medical Hypotheses* 59, no. 2 (August 2002): 170–79.

Sun Tzu. *The Art of War: Complete Texts and Commentaries*. Translated by the Denma Translation Group. Boulder, CO: Shambhala Classics, 2005.

Tai, Guangping, Michael Tai, and Min Zhao. "Electrically Stimulated Cell Migration and Its Contribution to Wound Healing." *Burns and Trauma* 6 (July 9, 2018): 20.

Temel, Jennifer S., Joseph A. Greer, Alona Muzikansky, Emily R. Gallagher, Sonal Admane, Vicki A. Jackson, Constance M. Dahlin, et al. "Early Palliative Care for Patients with Metastatic Non-Small-Cell Lung Cancer." *New England Journal of Medicine* 363, no. 8 (August 2010): 733–42.

Thomas, Frédéric, Thomas Madsen, Mathieu Giraudeau, Dorothée Misse, Rodrigo Hamede, Orsolya Vincze, François Renaud, Benjamin Roche, and Beata Ujvari. "Transmissible Cancer and the Evolution of Sex." *PLoS Biology* 17, no. 6 (June 2019): e3000275.

Tiede, Benjamin, and Yibin Kang. "From Milk to Malignancy: The Role of Mammary Stem Cells in Development, Pregnancy and Breast Cancer." *Cell Research* 21, no. 2 (February 2011): 245–57.

Tollis, Marc, Jooke Robbins, Andrew E. Webb, Lukas F. K. Kuderna, Aleah F. Caulin, Jacinda D. Garcia, Martine Bèrubè, et al. "Return to the Sea, Get Huge, Beat Cancer: An Analysis of Cetacean Genomes Including an Assembly for the Humpback Whale (Megaptera Novaeangliae)." *Molecular Biology and Evolution* 36, no. 8 (August 2019): 1746–63.

Topalian, Suzanne L., F. Stephen Hodi, Julie R. Brahmer, Scott N. Gettinger, David C. Smith, David F. McDermott, John D. Powderly, et al. "Safety, Activity, and Immune Correlates of Anti-PD-1 Antibody in Cancer." *New England Journal of Medicine* 366, no. 26 (June 2012): 2443–54.

Trigos, Anna S., Richard B. Pearson, Anthony T. Papenfuss, and David L. Goode. "Altered Interactions between Unicellular and Multicellular Genes Drive Hallmarks of Transformation in a Diverse Range of Solid Tumors." *Proceedings of the National Academy of Sciences of the United States of America* 114, no. 24 (June 2017): 6406–11.

Trivers, Robert L. "The Evolution of Reciprocal Altruism." *Quarterly Review of Biology* 46, no. 1 (March 1971): 35–57.

Turajlic, Samra, and Charles Swanton. "Metastasis as an Evolutionary Process." *Science* 352, no. 6282 (April 2016): 169–75.

Turner, Kristen M., Viraj Deshpande, Doruk Beyter, Tomoyuki Koga, Jessica Rusert, Catherine Lee, Bin Li, et al. "Extrachromosomal Oncogene Amplification

Drives Tumour Evolution and Genetic Heterogeneity." *Nature* 543 (February 2017): 122.

Ukraintseva, Svetlana V., Konstantin G. Arbeev, Igor Akushevich, Alexander Kulminski, Liubov Arbeeva, Irina Culminskaya, Lucy Akushevich, and Anatoli I. Yashin. "Trade-Offs between Cancer and Other Diseases: Do They Exist and Influence Longevity?" *Rejuvenation Research* 13, no. 4 (August 2010): 387–96.

Vaughan, Thomas L., Linda M. Dong, Patricia L. Blount, Kamran Ayub, Robert D. Odze, Carissa A. Sanchez, Peter S. Rabinovitch, and Brian J. Reid. "Non-Steroidal Anti-Inflammatory Drugs and Risk of Neoplastic Progression in Barrett's Oesophagus: A Prospective Study." *Lancet Oncology* 6, no. 12 (December 2005): 945–52. https://doi.org/10.1016/S1470-2045(05)70431-9.

Vousden, Karen H., and Xin Lu. "Live or Let Die: The Cell's Response to p53." *Nature Reviews Cancer* 2, no. 8 (August 2002): 594–604.

Waldman, Katy. "We're Finally Winning the Battle against the Phrase 'Battle with Cancer.'" *Slate*, July 30, 2015. https://slate.com/human-interest/2015/07/how-battle-with-cancer-is-being-replaced-by-journey-with-cancer.html.

Walsh, Justin T., Simon Garnier, and Timothy A. Linksvayer. "Ant Collective Behavior Is Heritable and Shaped by Selection." *bioRxiv* (March 2019): 567503.

Wang, Xu, Donald C. Miller, Rebecca Harman, Douglas F. Antczak, and Andrew G. Clark. "Paternally Expressed Genes Predominate in the Placenta." *Proceedings of the National Academy of Sciences of the United States of America* 110, no. 26 (June 2013): 10705–10.

Wang, Yu, Chenzhou Zhang, Nini Wang, Zhipeng Li, Rasmus Heller, Rong Liu, Yue Zhao, et al. "Genetic Basis of Ruminant Headgear and Rapid Antler Regeneration." *Science* 364, no. 6446 (June 2019). https://doi.org/10.1126/science.aav6335.

Wasielewski, H., J. Alcock, and A. Aktipis. "Resource Conflict and Cooperation between Human Host and Gut Microbiota: Implications for Nutrition and Health." *Annals of the New York Academy of Sciences* 1372, no. 1 (2016): 20–28.

Whisner, C., and A. Aktipis. "The Role of the Microbiome in Cancer Initiation and Progression: How Microbes and Cancer Cells Utilize Excess Energy and Promote One Another's Growth." *Current Nutrition Reports* 8, no. 1 (March 2019): 42–51.

White, Philip R., and Armin C. Braun. "A Cancerous Neoplasm of Plants: Autonomous Bacteria-Free Crown-Gall Tissue." *Cancer Research* 2, no. 9 (1942): 597–617.

Wirén, Sara, Christel Häggström, Hanno Ulmer, Jonas Manjer, Tone Bjørge, Gabriele Nagel, Dorthe Johansen, et al. "Pooled Cohort Study on Height and Risk of Cancer and Cancer Death." *Cancer Causes and Control* 25, no. 2 (February 2014): 151–59.

Witherow, Beth Ann, Gregory S. Roth, Mark A. Carrozza, Ronald W. Freyberg, Jonathan E. Kopke, Rita R. Alloway, Joseph F. Buell, et al. "The Israel Penn International Transplant Tumor Registry." *AMIA Annual Symposium Proceedings* (2003): 1053.

Wu, Shaoguang, Ki-Jong Rhee, Emilia Albesiano, Shervin Rabizadeh, Xinqun Wu, Hung-Rong Yen, David L. Huso, et al. "A Human Colonic Commensal Promotes

Colon Tumorigenesis via Activation of T Helper Type 17 T Cell Responses." *Nature Medicine* 15, no. 9 (September 2009): 1016–22.

Wynendaele, Evelien, Frederick Verbeke, Matthias D'Hondt, An Hendrix, Christophe Van De Wiele, Christian Burvenich, Kathelijne Peremans, Olivier De Wever, Marc Bracke, and Bart De Spiegeleer. "Crosstalk between the Microbiome and Cancer Cells by Quorum Sensing Peptides." *Peptides* 64 (February 2015): 40–48.

Yokoyama, Akira, Nobuyuki Kakiuchi, Tetsuichi Yoshizato, Yasuhito Nannya, Hiromichi Suzuki, Yasuhide Takeuchi, Yusuke Shiozawa, et al. "Age-Related Remodelling of Oesophageal Epithelia by Mutated Cancer Drivers." *Nature* 565, no. 7739 (January 2019): 312–17.

Zhang, Jingsong, Jessica J. Cunningham, Joel S. Brown, and Robert A. Gatenby. "Integrating Evolutionary Dynamics into Treatment of Metastatic Castrate-Resistant Prostate Cancer." *Nature Communications* 8, no. 1 (November 2017): 1816.

9 780691 212197